THE VAN NOSTRAND SCIENCE SERIES.

No. 19.—STRENGTH OF BEAMS UNDER TRANSVERSE LOADS. By Prof. W. Allan, author of "Theory of Arches." Second edition, revised.

No. 20.—BRIDGE AND TUNNEL CENTRES. By John B. McMaster, C.E. Second edition.

No. 21.—ᵒ ᵂᴵ. . VALVES. Second Editic · ᵤᵤchard ᴿ ᵇᵤₑₙ, C.E.

No. 22.—HIGH MASONRY ? ᵤIS. By E. ᵇᵤₑᵣₙₐₙ Goulᵢ, M. Am. Soc. C. ᵢᵢ,

No. 23.—THE FATIGUE OF METALS UNDER REPEATED STRAINS. With various Tables of Results and Experiments. From the German of Prof. Ludwig Spangenburgh, with a Preface by S. H. Shreve, A.M.

No. 24.—A PRACTICAL TREATISE ON THE TEETH OF WHEELS. By Prof. S. W. Robinson. Second edition, revised.

No. 25.—ON THE THEORY AND CALCULATION OF CONTINUOUS BRIDGES. By R. M. Wilcox, Ph. D.

No. 26.—PRACTICAL TREATISE ON THE PROPERTIES OF CONTINUOUS BRIDGES. By Charles Bender, C.E.

No. 27.—ON BOILER INCRUSTATION AND CORROSION. By F. J. Rowan. New Ed. Rev. by F. E. Idell.

No. 28.—TRANSMISSION OF POWER BY WIRE ROPES. Second edition. By Albert W. Stahl, U.S.N.

No. 29.—STEAM INJECTORS. Translated from the French of M. Leon Pochet.

No. 30.—TERRESTRIAL MAGNETISM AND THE MAGNETISM OF IRON VESSELS. By Prof. Fairman Rogers.

No. 31.—THE SANITARY CONDITION OF DWELLING-HOUSES IN TOWN AND COUNTRY. By George E. Waring, jun.

No. 32.—CABLE-MAKING FOR SUSPENSION BRIDGES. By W. Hildebrand, C.E.

No. 33.—MECHANICS OF VENTILATION. By George W. Rafter, C.E. New and Revised Edition.

No. 34.—FOUNDATIONS. By Prof. Jules Gaudard, C.E. Second edition. Translated from the French.

No. 35.—THE ANEROID BAROMETER: ITS CONSTRUCTION AND USE. Compiled by George W. Plympton. Eighth edition.

No. 36.—MATTER AND MOTION. By J. Clerk Maxwell, M.A. Second American edition.

No. 37.—GEOGRAPHICAL SURVEYING; ITS USES, METHODS, AND RESULTS. By Frank De Yeaux Carpenter, C.E.

No. 38.—MAXIMUM STRESSES IN FRAMED BRIDGES. By Prof. William Cain, A.M., C.E. New and revised edition.

THE VAN NOSTRAND SCIENCE SERIES.

ELECTROMAGNETS;

THEIR DESIGN,
AND CONSTRUCTION.

BY

A. N. MANSFIELD, S. B.

NEW YORK :
D. VAN NOSTRAND COMPANY.
23 MURRAY AND 27 WARREN STREETS.
1901.

Harry Hildenbrand [signature]

PREFACE.

THIS volume is the outgrowth of an attempt to revise Number 64 of "Van Nostrand's Science Series," *i. e.*, " Electromagnets, the Determination of the Elements of their Construction," by T. H. Du Moncel. It was found impossible to revise it without completely rewriting it, and it was deemed best to publish what follows as a new number in the "Science Series."

No claim is made for originality, but it has been attempted to collect, in a convenient form, formulæ and data which are essential to the design and construction of electromagnets for various purposes. We are indebted especially to the follow-

ing: S. P. Thompson, "The Electromagnet;" Du Bois, "The Magnetic Circuit in
Theory and Practice;" Jackson, "A Text
Book on Electromagnetism." Other
sources of information have been indicated in the foot-notes.

CONTENTS.

CHAPTER I.

ELECTROMAGNETS.

CHAPTER I.

1.—HISTORICAL.

In 1820 Oersted discovered that the neighborhood of a conductor conveying an electric current possessed magnetic properties. When a magnetic needle is brought near the conductor it is deflected, this effect being due to the presence of magnetic lines of force about the conductor.

In September of the same year, Arago described how he had communicated permanent magnetism to steel needles by laying them at right angles to a conductor conveying a current of electricity. Acting upon a suggestion of Ampere's, he greatly increased the magnetic effect by substituting a helix of wire for the straight

conductor, and placing the steel needle
in the axis of the helix.

William Sturgeon was the first to
devise an apparatus in which a soft iron
core surrounded by a coil of copper wire,
could be made to act as a magnet. Such
an apparatus is known as an *electromagnet.* Sturgeon described his first apparatus in a paper to the Society of Arts in
1825. He exhibited, with others, two
electromagnets, one of the horse-shoe
type, and one a straight bar.

Interest was aroused and much experi-
mental work was done by various investi-
gators. In the United States, Prof.
Henry, of Albany and Princeton, was an
ardent investigator from the year 1828.
Prof. Henry "was the first to actually
magnetize a piece of iron at a distance,
and to call attention to the fact of the ap-
plicability of his experiments to the telegraph." * He also "was the first to prove
by actual experiment that in order to
develop magnetic power at a distance, a

* Scientific Writings of Joseph Henry, Vol. II., p. 435.

galvanic battery of 'intensity' must be
employed to project the current through
the conductor, and that a magnet surrounded by many turns of one long wire
must be used to receive this current." *

About 1838, Joule of Manchester, be-
gan that work which added much to the
knowledge of the laws of the electromagnet. Joule obtained an insight into the
laws of the magnetic circuit and " found
it disadvantageous to increase the length
(of the iron core) beyond what is needful
for the winding of the covered wire." †
He did much work on the lifting power
of magnets and propounded the law that
the " attractive force of two electromagnets for one another is directly proportional to the square of the electric force
to which the iron is exposed ; or if E denote the electric current, W the length
of wire, and M the magnetic attraction,
$M = E^2 W^2$." ‡ The researches of Joule

* Scientific Writings of Joseph Henry, Vol. II., p. 435.
† Joules' Scientific Papers, Vol. I., p. 8.
‡ Ibid, p. 13.

may be said to end the first stage in the
development of the electromagnet.

Joule's notion of the magnetic circuit was not well received by physicists
and experimenters. Work was carried
on by many experimenters, among the
more prominent being Müller, Dub, Du
Moncel, Lenz, and Jacobi. They form-
ulated the results of their investigations
into empirical rules and laws, and little
was based on theory.

It is since 1880 that the magnetic circuit has been investigated mathematically
and the results of this work, combined
with experiment, have placed the laws of
magnetic circuits upon a sound and
practical footing. The work of Maxwell,
Rowland, Bosanquet, Hopkinson, Kapp,
Ewing, DuBois and others has all contributed towards this end. The consideration of the electromagnet is the
consideration of the laws of the magnetic
circuit, and this necessitates a knowledge
of the magnetic field and the magnetic
properties of iron. The study of the

13

magnetic field and the magnetic properties of iron has been largely contributed
to by Ewing, Steinmetz, Ebert and others.

2. MAGNETIC PROPERTIES.

The region surrounding a conductor
conveying an electric current possesses
magnetic properties, as is manifest by its
action upon magnetic substances, placed
within it. A *magnetic force* exists in the
region, and that region is called a *magnetic field*. The direction of the magnetic force in a magnetic field is indicated
by *lines of force*.

Lines of force, having the form of con-
centric circles with the conductor as a

Fig. 1.

center, surround a current conveying
conductor and lie in a plane perpendicular to the conductor. Direction is as-

cribed to these lines of force, and it is dependent upon the direction of the current. The directions are connected by the following relation: if one looks along a conductor in the direction in which the current is flowing, the positive direction of the lines of force is in the direction of motion of the hands of a watch. These directions are associated in Fig. 1.

The magnetic action of a current is increased, or concentrated, if instead of a linear conductor, the wire is given the form of a helix. The lines of force sur-

Fig. 2.

round the conductor and pass through the center of the helix as shown in Fig 2.

Such an arrangement is called a *solenoid*. A solenoid has the properties of a permanent magnet, i.e., has a magnetic field; attracts and repels magnetic bodies, etc.

The magnetic action of the solenoid is greatly increased if a core of soft iron is inserted into the helix; during the passage of a current through the coil, the core becomes strongly magnetic. The combination of coil and core forms an electromagnet.

The lines of force, or *magnetic flux*, generated by the current pass through the iron core, out at one end, through the air or surrounding medium and reenter the iron core at the other end. The end at which the lines of force are assumed to leave the core is called the north pole, and the end at which they are assumed to enter the core is called the south pole.

There are numerous rules for telling the polarity of an electromagnet when the direction of the current is known, one of the simplest being that of Jamieson, *i. e.* : place the right hand upon the winding so that the current is flowing in

the direction of from wrist to finger-tips;
the extended thumb points in the direction of the north pole.

3. CLASSIFICATION OF ELECTRO-MAGNETS.

There are three main divisions according to the functions of the electromagnet.

(a) *Attractive* electromagnets, namely, those in which the armature is
attracted, generally from a distance, when
a current flows through the coil.

(b) *Portative* electromagnets, namely, those which are used for temporary
adhesion or lifting power.

(c) *Field* electromagnets, namely,
those which are used for producing a
magnetic field of fixed strength, as in a
dynamo or motor.

Attractive electromagnets can be further subdivided into:

(a) *Polarized*, namely, those in
which the armature is a permanent magnet and is either attracted or repelled

by the core, depending upon the direction of the current in the coil.

(b) *Non-polarized*, namely, those in which the armature is of soft iron and the resultant action between the core and armature is always one of attraction.

4. FORMS OF ELECTROMAGNETS.

While there are many forms of electro-magnets they may be, for most purposes, classified as follows:

1. Bar electromagnets.
2. Horse-shoe electromagnets.
3. Iron-clad electromagnets.
4. Coil and plunger electromagnets.
5. Special forms of electromagnets.

Bar electromagnet. This is the simplest form of electromagnet and con-

Fig. 3.

sists of a straight iron bar surrounded by a coil of insulated wire as shown in Fig 3.

18

One of the principal uses of this form
is in the telephone receiver.

Horse-shoe electromagnet. This type
of electromagnet is very extensively
used, the iron core being of a U or horse-
shoe shape, as shown in Fig. 4.

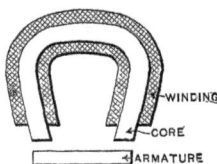

Fig. 4.

The core may be bent from a rod of
iron, or two separate iron-cores may be
connected by a third piece of iron, or the

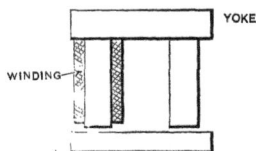

Fig. 5.

yoke. The winding may be uniformly
distributed over the core in the bent

form, may be collected on the two legs
or cores, may be collected on one core, or
may be collected on the yoke. When the
winding is collected on one core or leg, as
shown in Fig. 5, it is known as a *club-
foot* electromagnet.

Iron-clad electromagnet. This form
may be considered as a simple bar

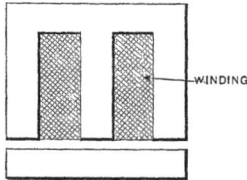

Fig. 6.

magnet having an iron shell outside of
the coil and attached to the core at one
end; or it may be considered as a form of
the club-foot magnet with one leg divided and surrounding the coil, as shown in
Fig. 6.

The iron shell forms a mechanical and
magnetic shield to the winding. The
free end of the core presents one pole and

the shell presents an outer annular pole
of the opposite polarity.

Coil and plunger electromagnet. This
is a simple coil or solenoid into which
a core may be sucked by magnetic action,
as shown in Fig. 7.

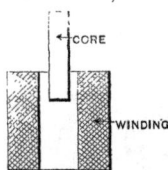

Fig. 7.

This form is used in many cases where
motion over a considerable distance is
desired.

Among the special forms of electromagnets the following may be mentioned:

Consequent pole electromagnet. When
an iron core is so wound that the current flows around the core in opposite
directions over two portions of the core,
like poles, say north poles, are formed
at the ends and a pole of opposite po-

larity, say a south pole, at the center. The central pole is known as a *consequent* pole.

The ordinary horse-shoe type of electromagnet is known as a *salient* pole magnet.

Circular electromagnet. If a circular piece of iron has a groove cut in its periphery and a magnetizing coil wound in this groove, one rim will be a north polar surface, and the other rim a south polar surface. This form of electromagnet is used in magnetic clutches.

CHAPTER II.

1. PROPERTIES OF THE MAGNETIC CIRCUIT.

Unless otherwise specified, a closed ring of soft iron uniformly wound will be considered in what follows.

When a current flows through the winding of an electromagnet, a magnetic force is generated which sets up and maintains a magnetic state or stress in the core. The core is said to be traversed by a *magnetic flux* or by *magnetic induction*. The *intensity* of the magnetic induction, or the *flux density* is measured by the number of lines of force which pass through a unit (square centimeter) area in a direction at right angles to the lines of force, and it is generally denoted by the letter B.

The force which produces this magnetic induction is called the *magnetomotive force* and the magnetomotive force per unit length (cm.) of magnetic circuit is called the *magnetizing force ;* the former being generally abbreviated to M.M.F.,or simply M, and the latter being generally denoted by the letter H.

If N represents the total number of spirals, or convolutions, in the winding surrounding the ring, and I the amperes flowing, the product NI is called the *number of ampere-turns*. The relation which exists between ampere-turns and M is

$$M = \frac{4\pi}{10} NI = 1.257 \, NI. \qquad (1)$$

If l is the length (mean perimeter) of the ring in cms.

$$H = \frac{M}{l} = 1.257 \, \frac{NI}{l}. \qquad (2)$$

If $n = \frac{N}{l}$, i. e., n is the number of convolutions per centimeter length,

$$H = 1.257 \, nI \qquad (3)$$

= 1.257 times the ampere-turns per centimeter.

If l is given in inches

H = 0.495 times the ampere-turns per inch.

24

The density of the magnetic flux,
which is produced by a given M.M.F.,
depends upon the configuration and material of the core. The flux density produced in a magnetic body is many times
greater than that produced in a non-
magnetic body of the same dimensions.
That property of the core which limits
the flux density is called the *magnetic reluctivity* of the material. Air is taken
as the standard of comparison, and the
reluctivity of a cubic centimeter of air is
taken as unity. The reluctivity of all
non-magnetic substances is practically
unity, and the reluctivity of magnetic
substances is very small as compared to
that of air. The reluctivity of a body
multiplied by its length and divided by
its cross-sectional area is called its *reluctance.*

The following relations exist between
the several magnetic quantities referred
to above, where *l* is the mean perimeter
of the ring (cms.) and *s* its cross-sectional area (sq. cms).

First $\qquad$ F = Bs $\qquad$ (4)

that is, the total flux or induction equals the flux density multiplied by the cross-sectional area.

Second. $\qquad$ M = Hl. $\qquad$ (5)

that is, the magneto-motive force equals the product of the magnetizing force aud the length of the magnetic circuit.

Third. $\qquad$ B = μH $\qquad$ (6)

that is, the magnetic induction equals the magnetizing force multiplied by a factor μ, which is called the *permeability* of the material. The permeability is

$$\mu = \frac{B}{H}. \qquad (7)$$

μ is dependent upon the quality of the iron and the extent of the magnetization.

A curve which shows graphically the relation between the induction B and the magnetizing force H is called a *curve of magnetization*, and has the general shape shown in Fig. 8.

The curve which shows the relation

between the permeability μ and the induction B is called the *permeability curve,*

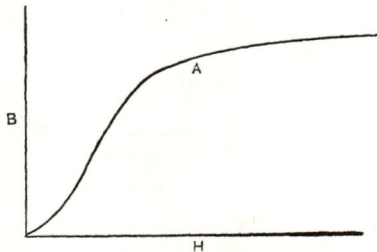

Fig. 8.

and has the general shape shown in

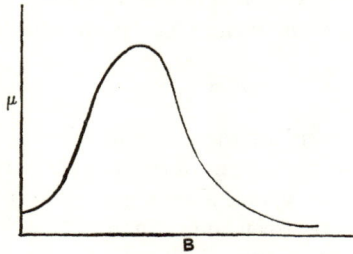

Fig. 9.

Fig. 9. This curve may be constructed from the curve of magnetization.

For small magnetizing forces the permeability is small; as H increases μ increases and therefore B; after reaching a certain point the permeability begins to decrease, and this corresponds to a bend in the magnetization curve as at A, Fig. 8. At this stage the iron is said to be approaching saturation. A knowledge of the relation between B, H, and μ for different qualities of iron is essential in the design of all electromagnetic apparatus.

Fourth.
$$S = \frac{1}{\mu} = \frac{H}{B} \cdot \qquad (8)$$

that is, the *magnetic reluctivity* is equal to the reciprocal of the permeability, or to the magnetizing force divided by the induction.

Fifth.
$$P = \frac{\mu s}{l} \cdot \qquad (9)$$

that is, the *permeance*, P, equals the permeability multiplied by the cross-sectional area of the magnetic circuit, and the product divided by the length of the circuit.

Sixth. The reciprocal of the permeance

$$X = \frac{1}{P} = \frac{l}{\mu s} = \frac{Sl}{s}. \tag{10}$$

that is, the *reluctance*, X, equals the reluctivity multiplied by the length of the circuit and divided by the cross-sectional area of the circuit. The reluctivity is the reluctance per unit volume.

From a consideration of the above equations the following relations hold:

$$B = \mu H. \tag{6}$$

Multiplying both sides of the equation by $\frac{s}{l}$ and transforming

$$Bs = \frac{\mu s}{l} Hl.$$

or

$$F = MP = \frac{M}{X}. \tag{11}$$

i. e. the total flux is equal to the magneto-motive force multiplied by the permeance, or divided by the reluctance. This proposition for the magnetic circuit is

analogous to Ohm's Law for the electric
circuit; this difference must however be
noted, that the electrical resistance is absolutely independent of the strength of
of the current, depending upon the material of the conductor, its temperature
and its geometrical dimensions; while the
magnetic resistance or reluctance of the
magnetic circuit is dependent upon the
magnetic intensity, being low for low
intensities, increasing to a maximum and
then decreasing and approaching unity
when the magnetizing force is high.
The analogy of Ohm's Law to magnetic
circuits is only an artifice to facilitate in-
vestigation.

The above relation of the magnetic
circuit may be written

$$\text{Magnetic Flux} = \frac{\text{Magnetomotive Force}}{\text{Reluctance}}.$$

$$(12)$$

Names have been given to the magnetic
terms analogous to the electric terms.
One ampere turn is frequently used as

the unit of M.M.F. Another unit called the *gilbert*, is in use, one gilbert being equal to $\frac{10}{4\pi}$ (= 0.796) ampere turns, or one ampere turn is equal to 1.257 gilberts. The unit of reluctance is called the *oersted*, and is equal to the reluctance which is offered to the passage of magnetic flux, by a cubic centimeter of air when measured between parallel faces. The unit of magnetic flux is called the *weber*, and is the amount of flux which would pass through a magnetic circuit whose reluctance is one oersted, when the M.M.F. acting is one gilbert. The law of the magnetic circuit can then be written

$$\text{Webers} = \frac{\text{Gilberts}}{\text{Oersteds}}, \tag{13}$$

The magnetic density, i. e. the flux per unit area is called the *gauss*, and is a density of one weber passing perpendicularly through an area of one square centimeter.

Substituting the value of H as given in (2) we have

$$B = \mu H = \frac{4\pi}{10}\mu \frac{NI}{l} = 1.257 \frac{\mu NI}{l}.$$

Also

$$F = \frac{M}{X} = \frac{Hl}{\dfrac{l}{\mu s}} = \frac{\dfrac{4\pi}{10}NI}{\dfrac{l}{\mu s}}, \qquad (14)$$

or

$$F = \frac{1.257\ NI}{\dfrac{l}{\mu s}}. \qquad (15)$$

which may be called the fundamental law of the simple magnetic circuit.

Transposing, $NI = 0.795\,F \times \dfrac{l}{\mu s}.$ (16)

If inch measure is used,

$$NI = 0.3132F \times \frac{l^1}{\mu s^1}, \qquad (17)$$

the prime mark being added to denote inch measure.

When the magnetic circuit consists of several parts, such as cores, yoke, armature and air gaps; or of parts with different cross-sectional areas and permeabilities, whose separate reluctances are x_1, x_2, x_3, etc., the total reluctance of the circuit is

$$X = x_1 + x_2 + x_3 + \ldots \quad (18)$$

The reluctance of a circuit is equal to the sum of the reluctances of the component parts, as in a series electrical circuit the total resistance is the sum of the separate resistances.

Let μ_1, μ_2, μ_3, etc., l_1, l_2, l_3, etc., s_1, s_2, s_3 etc. represent respectively the permeabilities, lengths, and cross-sectional areas, of the several component parts of a magnetic circuit, then

$$X = \frac{l_1}{\mu_1 s_1} + \frac{l_2}{\mu_2 s_2} + \frac{l_3}{\mu_3 s_3} +, \text{ etc.} \quad (19)$$

$$\text{and } F = \frac{1.257 NI}{\dfrac{l_1}{\mu_1 s_1} + \dfrac{l_2}{\mu_2 s_2} + \dfrac{l_3}{\mu_3 s_3} + \ldots} \quad (20)$$

Transposing have

$$NI = \frac{F \times \left\{ \dfrac{l_1}{\mu_1 s_1} + \dfrac{l_2}{\mu_2 s_2} + \dfrac{l_3}{\mu_3 s_3} + .. \right\}}{1.257} \quad (21)$$

that is

$$NI = 0.795F \left\{ \frac{l_1}{\mu_1 s_1} + \frac{l_2}{\mu_2 s_2} + \frac{l_3}{\mu_3 s_3} + .. \right\} \quad (22)$$

when the dimensions of the circuit are expressed in centimeters.

If the unit of measurement is the inch, the formula becomes

$$NI = 0.3132F \left\{ \frac{l_1^{\;1}}{\mu_1 s_1^{\;1}} + \frac{l_2^{\;1}}{\mu_2 s_2^{\;1}} + \frac{l_3^{\;1}}{\mu_3 s_3^{\;1}} + .. \right\} \quad (23)$$

What amounts to the same as is expressed by formulae (22) or (23), is to apply (16) or (17) to each separate portion of the magnetic circuit and determine the ampere-turns required to maintain the flux F through that portion of the circuit. The sum of the ampere-turns thus determined gives the total ampere-

turns required and should agree with the
number obtained by (22) or (23).

Effect of air gap in magnetic circuit.—
If air or any non-magnetic material is in-
cluded in the magnetic circuit, the reluctance of the circuit is increased many
times. The permeability of air being
taken as unity the relation $B = H$, i.e.,
induction is equal to magnetizing force,
holds for that portion of the circuit made
up of air. The ampere-turns required
per unit length to produce a given induction across an air gap are in centimeter measure

$$NI \text{ (per cm.)} = 0.796 \, B \text{ (per sq. cm.)} \quad (24)$$

or, in inch measure,

$$NI \text{ (per inch)} = 0.313 \, B' \text{ (per sq. inch)}. \quad (25)$$

In addition to increasing the reluctance of the circuit, an air gap intro-
duces leakage and a demagnetizing action
due to the influence of the poles induced
at the ends of the core.

Effects of Joints in Magnetic Circuit.— Joints or cracks, no matter how fine, exert an influence upon the magnetic circuit by increasing its reluctance and leakage. If a magnetic circuit contains a joint, equation (15) takes the form

$$F = \frac{1.257NI}{\dfrac{l}{\mu s} + \dfrac{a}{s}} = \frac{1.257NI}{\dfrac{1}{\mu s}(l + a\mu)} \qquad (26)$$

where a is the equivalent air gap for the joint. The length of the magnetic circuit has been increased by an amount equivalent to $a\mu$. Ewing and Low found that the equivalent air gap for two wrought-iron bars was about 0.003 of a centimeter, or 0.0012 of an inch. The effect of the joint is most noticeable for low magnetizations; for when the iron approaches saturation the component parts of the circuit are strongly attracted, and the resultant pressure reduces the effect of the joint.

In jointing a magnet frame, it is of the utmost importance that the abutting faces be accurately finished, so as to make the joint as close as possible. As noticed above, if the circuit is worked near the saturation point, the influence of the joint is not so pronounced. Where joints occur between portions of the same circuit, and these portions are of different permeabilities, the cross-sectional area of the joint, or the total area perpendicular to the flux should be equal to the cross-sectional area of that portion of the circuit having the lowest permeability.

Magnetic Leakage.— In an electro-magnet the magnetic circuit can be considered as made up of three parts: the cores and yoke, the armature, and the air space, or joint, between armature and cores. When the armature is in contact with the cores, the reluctance of the circuit is low and practically made up of the iron reluctance. When there is an air gap the reluctance is greatly increased, and is made up mostly of the gap reluctance. In this case there is a considerable leakage of lines of force across from core to core, as indicated in Fig. 10.

The leakage paths are in parallel with the useful path for lines of force through the cores and armature. The number of lines of force in each path varies inversely as the reluctance and directly as

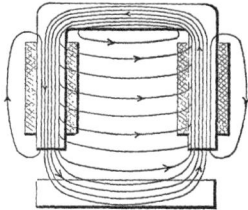

Fig. 10.

the magnetomotive force which is acting. The laws of parallel circuits can be applied to magnetic circuits, as they are applied to electric circuits, but the dimensions and conditions of the former are not as definite as in the case of the latter.

The reluctance between surfaces of

different shapes, sizes and inclinations
can be deduced, but the one of principal
import in electromagnet design is that
between parallel cylindrical surfaces. If
the diameter of the cylinders is d, the
distances between their centers is b, and
their length is l, the air reluctance* between the cylinders is

$$X = \frac{0.737 \ \log._{10} \frac{a_1}{a_2}}{l} \qquad (27)$$

where $\quad \dfrac{a_1}{a_2} = \dfrac{d}{b - \sqrt{b^2 - d^2}} \qquad (28)$

The numerical value of $\dfrac{a_1}{a_2}$ is constant
for all dimensions as long as the ratio
$\dfrac{b}{d}$ is constant.

The following table† gives the mag-

---

* Jackson, " Electro-Magnetism and the Construction
of Dynamos," p. 131.
† Taken, by permission of the publishers, from Professor
Dugald C. Jackson's Text-Book on Electromagnetism and
the Construction of Dynamos, copyright by The Macmillan
Company, 1893.

netic reluctance in c.g.s units between
unit lengths of two equal parallel cylinders surrounded by air and having various values of the ratio $\dfrac{b}{d}$.

$\dfrac{b}{d}$	X per cm.	$\dfrac{b}{d}$	X per cm.	$\dfrac{b}{d}$	X per cm.
1.25	0.19	4.0	0.655	7.5	0.86
1.50	0.30	4.5	0.67	8.0	0.88
1.75	0.337	5.0	0.73	8.5	0.90
2.00	0.42	5.5	0.76	9.0	0.92
2.50	0.50	6.0	0.79	9.5	0.94
3.00	0.556	6.5	0.815	10.0	0.96
3.50	0.61	7.0	0.84		

If the dimensions of the cylinders
are given in inches, X per cm. must be
divided by 2.54 to give X per inch.
Hence, to find the reluctance between
any two cylinders, find the value of $\dfrac{b}{d}$
from the dimensions, and take the proper
value of X per cm. or inch, and divide
by the length of the cylinder in cms. or
inches.

Consider an electromagnet in which
the magnetizing coil is wound upon the
keeper, or yoke, as in Fig. 11.

40

In this case the magnetomotive force may be considered as acting between the leakage surfaces, and also across the air gap between pole pieces and armature.

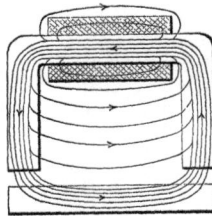

Fig. 11.

The following approximate relations hold. The total leakage lines are

$$N_1 = \frac{M}{X_1} = MP_1 ,$$

where X_2 and P_2 are the reluctance and permeance of all the leakage lines in parallel. The useful lines are

$$N_a = \frac{M}{X_a} = MP_a ,$$

where X_a and P_a refer to the circuit

made up of cores, air gap, yoke and armature. The total number of lines of force generated is

$$N_t = N_1 + N_a,$$

and the ratio

$$\frac{N_t}{N_a} = \frac{X_a + X_1}{X_1} = \frac{P_a + P_1}{P_a}.$$

The ratio $\frac{N_t}{N_a}$ is denoted by the letter v, and is called the leakage coefficient. When the reluctance X_a is small, the leakage coefficient is about unity.

When the magnetizing coils are divided between the two legs of an electromagnet, as shown in Fig. 10, the leakage lines cannot be considered as directly proportional to P_1. The difference of magnetic pressure between A and B is approximately zero, and that between C and D approximately equal to M, and hence the average difference of magnetic pressure between the two legs is $\frac{M}{2}$.

Hence, $N_1 = \dfrac{M}{2X_1} = \dfrac{M}{2} P_1.$

From this consideration the value of v may be found.

If the reluctance of the cores is not small, the magnetic pressure acting between C and D cannot be considered as M, and the problem is complicated by corrections to be applied for this effect.

From the above consideration it is seen that in general the more uniformly the winding is distributed over the magnetic circuit the less the leakage.

Roundness and evenness of the magnetic circuit, avoidance of sharp corners and abrupt turns, all tend to reduce leakage to a minimum.

Hysteresis.—When a piece of iron is subjected to a series of gradually increasing magnetizing forces, beginning at zero, the resultant induction increases until the iron becomes saturated. The relation between B and H can be plotted as a curve, which has the general form shown in Fig. 8. If now the magnetizing force is gradually decreased from this maximum value to zero, it will be found that the resultant flux does not have the same value for corresponding values of H, in the increasing and decreasing series of values of H; being greater in the decreasing series. The magnetic flux in a piece of iron, or other magnetic material, depends upon the previous values of the M. M. F. as well as the present ones. The flux lags behind the magnetizing force, and this lagging is known as *hysteresis*.

When the iron is subjected to a complete cycle of changes of magnetizing forces, that is, from zero to a positive maximum, then back to zero, then to a negative maximum, and back to zero, and the relation between B and H is plotted, the flux lags behind the M. M. F., and produces a closed loop, as shown in Fig. 12.

This loop is known as the *hysteresis loop*. The area enclosed by the hysteresis loop is a measure of the energy lost in each unit volume per cycle of change ; this loss of energy is made manifest in the form of heat in the iron. The hysteresis action then produces a loss of energy by heating, and in the

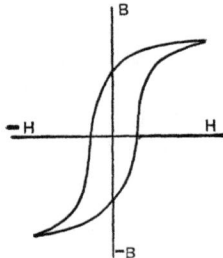

Fig. 12.

case of alternate current apparatus may produce serious heating.

The energy loss due to heating is given for a wide range of flux densities by the following formula, due to Steinmetz.

$$w = f_n \, B_{max}^{1.6}, \tag{29}$$

where

$w =$ watts lost per cubic centimeter of iron.

f = number of cycles per second.
B = maximum induction per sq. cm.
and
η = a constant varying from 0.002
for soft iron to 0.004 or 0.0045
for transformer irons.

Annealing, while it promotes homo-
geneity in the iron and increases the permeability of the hysteresis, also increases
the constant if it be originally very small.

Retentiveness.—When a piece of iron
has been magnetized and the magnetizing force withdrawn, the magnetization
does not entirely disappear, but a greater
or less amount remains, depending upon
the quality of the iron. That portion of
the magnetization which remains is called
residual magnetization, and that property
of retaining magnetization is known as
the *magnetic retentivity* or *retentiveness*
of the iron. The iron acts as if an internal M.M.F. was present and tended
to maintain the magnetic flux. This
force is called the *coercive force.*

Soft irons have a small retentiveness, while that of hard steel is high.

Retentiveness produces a sticking of the armature in electromagnetic mechanisms, but when the armature is detached from the pole pieces the retentiveness practically disappears.

From the above, it is seen that the magnetic properties of iron play an important part in the design and efficiency of electromagnets, as the reluctance of the circuit depends upon the quality of the iron used.

In general, it may be said that the magnetic properties of iron and steel depend upon the percentage of carbon present, and to a less degree upon the percentage of sulphur, phosphorus, manganese, silicon, and other impurities.

The qualities of several different grades of iron are as follows:

Cast Iron.—Cast iron is hard and comparatively brittle. It contains considerable carbon, part of it in a combined state and the rest in a free state. Irons containing more than 0.8 per cent of combined carbon are of a low magnetic permeability, and those having less than 0.3 per cent are of a high permeability. The combined carbon should be kept as low as possible, while the free carbon may vary from 2 to 3 per cent without having any marked effect on the permeability. Cast iron has a low cost, and can easily be obtained in any desired shape.

Wrought Iron.—Wrought iron is comparatively soft, malleable, and ductile. It contains a very small percentage of carbon. Wrought iron shows uniformly high permeability. It is hard to produce cheap in any but the simplest shapes.

Cast Steel.—This has a very small percentage of combined carbon and no free carbon. Good cast steel should not contain more than 0.25 per cent of carbon. It also contains phosphorus, silicon, manganese and sulphur as impurities.

In exact work, the permeability should be obtained for a sample of each lot of

48

iron used. The following table, taken
from Wiener's "Dynamo Electric Machines," p. 311, gives the general rela-
tionship between induction and permeability for several grades of iron. The
results given are the average permeabilities of a large number of tests of the various grades.

PERMEABILITIES OF DIFFERENT KINDS OF IRON AT VARIOUS MAGNETIZATIONS.

DENSITY OF MAGNETIZATION.		PERMEABILITY μ.			
Lines per Sq. Inch. B_1	Lines per Sq. cm. B	Annealed Wrought Iron.	Comm'l Wrought Iron.	Gray Cast Iron.	Ordin'y Cast Iron.
20,000	3,100	2,600	1,800	850	650
25,000	3,875	2,900	2,000	800	700
30,000	4,650	3,000	2,100	600	770
35,000	5,425	2,950	2,150	400	800
40,000	6,200	2,900	2,130	250	770
45,000	6,975	2,800	2,100	140	730
50,000	7,750	2,650	,2,050	110	700
55,000	8,525	2,500	1,980	90	600
60,000	9,300	2,300	1,850	70	500
65,000	10,100	2,100	1,700	50	450
70,000	10,850	1,800	1,550	35	350
75,000	11,650	1,500	1,400	25	250
80,000	12,400	1,200	1,250	20	200
85,000	13,200	1,000	1,100	15	150
90,000	14,000	800	900	12	100
95,000	14,750	530	680	10	70
100,000	15,500	360	500	9	50
105,000	16,300	260	360		
110,000	17,400	180	260		
115,000	17,800	120	190		
120,000	18,600	80	150		
125,000	19,400	50	120		
130,000	20,150	30	100		
135,000	20,900	20	85		
140,000	21,700	15	75		

Limits of Magnetization.—Iron is said
to be saturated when a very material increase in the magnetizing force does not
appreciably increase the magnetic flux.
This limit of saturation in different qualities of iron is reached at the values given
in the following table:*

. VALUE OF B.

	Per Sq. cm.	Per Sq. Inch.
Wrought Iron........	20,200	130,000
Cast Steel...........	19,800	127,500
Mitis Iron...........	19,000	122,500
Cast Iron (ordinary)...	12,000	77,500

The practical working densities are
about two-thirds of the densities given
in the above table.

Magnetic Testing.—The relation between magnetizing force, permeability,
and induction may be obtained in numerous ways. The methods may be
broadly divided into four general classes:

    (1) Magnetometric.
    (2) Balance.
    (3) Inductive or Ballistic.
    (4) Tractive.

---

* Wiener, Dyn. Elec. Machinery, p. 313, Table LXXVI.

For a general description of the methods, the reader is referred to Ewing, "Induction in Iron and other Metals;" Du Bois, "The Magnetic Circuit in Theory and Practice;" also a paper by Ewing in the "Proc. Inst. Civ. Eng.," Vol. CXXVI (1896).

Two methods which can be generally used are the ballistic and tractive methods.

Ballistic Method.—This method may be subdivided, in accordance with the form which the test piece of iron takes, into the Ring Method and the Bar and Yoke Method.

The method depends upon the measurement of the transient electric pressure induced in a small test coil, wound around the test piece, when the induction in the test piece is changed. A sample of the iron to be tested is inserted in a magnetic circuit whose magnetizing force is accurately known. A small test coil of known dimensions surrounds the test piece and is connected to a ballistic galvanometer. The induction passing through the coil is caused to vary by making, breaking, or reversing the magnetizing current; or the test coil is suddenly removed to a position where the intensity of the field may be neglected. The value of the induction may be obtained from the throw of the ballistic galvanometer needle, when the constant of the galvanometer is known. This constant may be obtained in several ways, as explained in the above references, or in Jackson's "Electro Magnetism and the Construction of the Dynamo." The ballistic method may be used for measuring fields of any intensity.

Ring Method.—The general arrangement of the apparatus in the ring method is shown in Fig. 13.

A is a ring made of the iron to be tested, and is wound with a magnetizing coil, P, which is in series with a source of current, D, through an adjustable resistance, such as a lamp bank. A double throw switch, Sw, is connected into this

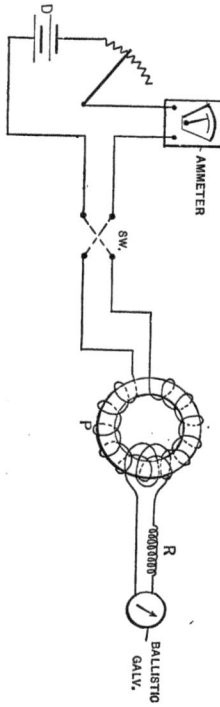

Fig. 13.

D

AMMETER

SW.

P

R

BALLISTIO
GALV.

primary circuit, so that the direction of current through the coil may be reversed. A secondary winding is connected to a ballistic galvanometer, through an adjustable resistance, R. The magnetizing force is

$$H = \frac{1.25 \; NI}{10 \, l}.$$

When the primary circuit is made or broken, a deflection is produced in the ballistic galvanometer proportional to the magnetic flux through S ; i.e., $B = kd$, where d is the deflection. The constant, k, of the ballistic galvanometer has to be obtained by calibration.

The advantages of this method are that it only requires a ring of iron, which may be cast from the desired lot or " run "; there is no finishing of the piece to obtain good joints; and the apparatus is much easier to set up and work than the yoke method.

A cylindrical specimen may be used if the ratio of $\dfrac{\text{length}}{\text{diameter}}$ is great, and the

secondary is wound over a short portion
of the bar.

Tractive Method.—This method, also
known as the "permeameter method,"
is described in many text-books. The
general arrangement is shown in Fig. 14.

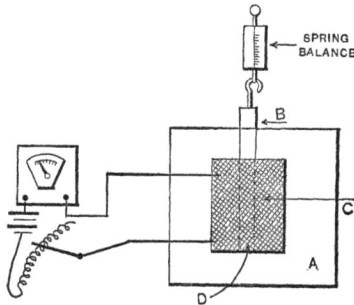

Fig. 14.

A is a rectangular block of soft iron,
cut out to receive a magnetizing coil, C,
within which is a thin brass tube. The
test piece, B, is placed in the tube and
makes contact with the block at D, which
joint is very carefully surfaced. When

a current passes through the magnetizing coil, the test piece adheres tightly to the block at D. The force which is required to detach the test piece is measured by the spring balance, S, which is carefully graduated and provided with an automatic catch so that the index sets at the highest reading. The force which is required to detach the test piece is a function of the induction across the joint, D.

When the yoke is large in relation to the test rod, the reluctance of the circuit is practically that of the test piece.

$$M = \frac{4\pi NI}{10} = FX = \frac{FL}{\mu A} = \frac{BL}{\mu} = HL.$$

$$\therefore \qquad H = \frac{4\pi NI}{10 L},$$

where L equals the mean length of the lines of force in the test piece; i.e., the sum of the length of the test piece and the opening in the block, divided by two. If the yoke reluctance cannot be neglected, then a connection has to be applied, for which see treatises on magnetic properties of iron and testing.

As will be noted later, the following relation exists between B, H, the cross-sectional area A, and the force required to detach the test piece.

$$B = 157 \sqrt{\frac{W}{A}} + H, \qquad (30)$$

where W is in grammes, and A is in sq. cms.,

or, $$B = 1317 \sqrt{\frac{P}{A_1}} + H, \qquad (31)$$

where P is in pounds, and A_1 is in sq. inches.

While this method is subject to several sources of error, it presents a method which is easy of application, and which will give results sufficiently accurate in magnet design. The joint, D, is in an unfavorable position, and Ewing has proposed that the joint should be in the middle of the core, or bar.

Instead of using the above form of

test piece, the iron may be formed into
two semi-circular rings, around which
the magnetizing turns are placed, and
then the force necéssary to separate them
is measured. The first method, however,
is more accurate.

CHAPTER III.

THE materials which enter into the construction of an electromagnet are those of the core, the conductor, the insulation and the frame.

Conductor material.—Copper is the only material which is commercially used as a conductor. It should be soft and have a conductivity of at least 98% of that of pure soft copper. The cross-section of the conductor is not necessarily circular ; in many cases wires of a square or rectangular cross-section are used, and these shapes may either be solid or made up of ribbons, and bound together. In cases where heavy currents are to be carried, a stranded conductor can often be advantageously used, on account of the greater facility with which it can be handled and wound. For equal diameters, cables or strands have a conducting area of from 20 to 25 per cent. less than a solid circular conductor.

Insulating material.—Insulating material is employed to prevent leakage and short circuits between adjacent turns and layers, or contact with the frame. The principal properties which an insulating material should possess are, (1) a high insulation resistance, (2) it should not absorb moisture, (3) it should withstand high temperatures without injury, (4) it should not be punctured by high voltages, and (5) it should be chemically inert. It is impossible to find any one substance possessing all of the above properties in combination, and thus the selection of the insulating material has to be judged by the case in hand.

Wire insulation.—The insulation of the wire usually consists of a cotton or silk covering, single, double, or triple, as the case may demand; cotton being the more commonly used. On large sized wires a cotton braid is often used in place of a double cotton covering. For small work such as telephone, telegraph, bell or signal work, single cotton, and on the smaller sizes of wire single or double silk covering is used. If the winding is to be moisture proof it should be thoroughly soaked in shellac and dried, or it may be immersed in hot paraffine or beeswax. Stranded and rectangular windings are generally covered with a cotton tape and shellaced after winding.

Layer insulation.—In small electromagnets there is little need for insulation between the layers. Where high voltages are employed, oiled linen, vulcanized fibre, or mica should be used. If the core and heads of the electromagnet are of metal they should be covered with some good insulating paint, or the core covered with hard rubber or vulcanized fibre. Mica, vulcanized fibre, oiled or paraffined paper, etc., should be placed between the frame heads and the winding.

When a magnet is to be highly heated and there is danger of the insulation becoming carbonized, asbestos or special fire-proof paints should be used.

Frame.—The material of the coil frame depends upon the character of the magnet, according to the special requirements of the case, as to mechanical strength, cost, etc. Wood, hard rubber, vulcanized fibre, etc., can be used in small magnets, while brass may be used in large ones.

Core material.—Iron and steel are solely used as a core material and their properties have been discussed in another place.

Form of cross-section of magnet core.—The best form of magnet core is that which possesses the smallest circumference for a given area. The following table gives the circumferences for unit areas and the relative circumferences for equal areas :

63

Description.	Circumference for Unit Area.	Relative Circumfer'ce Circle = 1.
Circle.	3.545	1.00
Square.	4.000	1.13
Rectangle 1:2	4.243	1.20
Rectangle 1:3.	4.620	1.30
Ellipse 1:2.	3.87	1.09
Ellipse 1:3.	4.35	1.28
Oval 1 sq. 2 semicircles.	3.85	1.08
Oval 2 sq. 2 semicircles.	4.28	1.21
2 circles, wire around both, as indicated by dotted lines.	4.10	1.13

From this table it is seen that the circle gives the most economical form of cross-section of core, as less copper is required per convolution of conductor. Circular cores are also of advantage, as leakage from core to core is, for equal mean-distances apart, proportional to the surface of the core, and the circular core has the minimum surface for equal areas. In circular cores all sharp edges are avoided and leakage is minimized.

64

Thickness of wire insulation.—The
thickness of insulation employed upon
wires varies with the manufacturer and
no fixed value can be given to cover all
cases. The following table represents the
practice of several large manufacturers.
To determine the *diameter* of insulated
wire add to the *diameter* of the bare wire:

B. & S. Guage.	For Cotton.		For Silk.	
	Single,	Double.	Single.	Double.
0 to 10	7 mils.	14 mils,		
10 to 18	5 "	10 "		
18 up.	4 "	8 "	2 mils.	4 mils.

The following table constructed from
Tables XXVI and XCII in Wiener's
"Dynamo Electric Machines" gives the
ratio of weight between covered and bare
wires. It should be noticed that the
thickness of insulation given by Wiener
is larger than that given in the above
table.

TABLE GIVING THICKNESS AND WEIGHT OF INSULATION ON WIRES.

GAUGE OF WIRE.		Diameter of Wire Bare.	SINGLE COTTON INSULATION.			DOUBLE COTTON INSULATION.	
B. W. S.	B. & S.		Thickness of Insulation, inches.	Weight of Covered Wire per pound of Bare Wire.	Ratio of Copper to total volume of Coil.	Thickness of Insulation, inches.	Weight of Covered Wire per pound of Bare Wire.
1	...	.300	....	.....	....	.020	1.0228
...	1	.289	....	.....	....	.020	1.0232
2	...	.284	....	.....	....	.020	1.0233
3	...	.259	....	.....	....	.020	1.024
...	2	.258	....	.....	....	.020	1.024
4	...	.238	....	.....	....	.020	1.025
...	3	.229	....	.....	....	.020	1.0255
5	...	.220	....	.....	....	.020	1.0265
...	4	.204	.012	1.022	.702	.020	1.0285
6	...	.203	.012	1.022	....	.020	1.0286
...	5	.182	.012	1.0227	.69	.018	1.0287
7	...	.180	.012	1.0228	....	.018	1.029
8	...	.165	.012	1.0233	.683	.018	1.032
...	6	.162	.010	1.0224	.697	.018	1.0325
9	...	.148	.010	1.023	.688	.016	1.0315
...	7	.144	.010	1.0232	....	.016	1.0325
10	...	.134	.010	1.0236	.682	.016	1.0355
...	8	.1285	.010	1.024	....	.016	1.0375
11	...	.120	.010	1.025	.669	.016	1.041
...	9	.1144	.010	1.0255	....	.016	1.0435
12	...	.109	.010	1.0266	.66	.016	1.046
...	10	.102	.010	1.0285	.65	.016	1.05
13	...	.095	.010	1.031	.644	.016	1.0555
...	11	.091	.010	1.0325	.637	.016	1.0585
14	...	.083	.007	1.025	....	.016	1.066
...	12	.081	.007	1.0254	.665	.016	1.068
15	13	.072	.007	1.028	.65	.016	1.078

TABLE GIVING THICKNESS AND WEIGHT OF INSULATION ON WIRES—*Continued.*

GAUGE OF WIRE.		Diameter of Wire Bare.	SINGLE COTTON INSULATION.			DOUBLE COTTON INSULATION.	
B. W. S.	B. & S.		Thickness of Insulation, inches.	Weight of Covered Wire per pound of Bare Wire.	Ratio of Copper to total volume of Coil.	Thickness of Insulation, inches.	Weight of Covered Wire per pound of Bare Wire.
16	...	.065	.007	1.0315	.64	.016	1.086
...	14	.064	.007	1.0325	....	.016	1.088
17	...	.058	.007	1.086	.625	.014	1.086
...	15	.057	.007	1.037	....	.014	1.086
...	16	.051	.007	1.042	.607	.014	1.096
18	...	.049	.007	1.044	....	.014	1.098
...	17	.045	.005	1.0325	.637	.012	1.093
19	...	.042	.005	1.0355	.627	.012	1.098
...	18	.040	.005	1.0375	.628	.012	1.101
...	19	.036	.005	1.043	.607	.005*	1.056
20	...	.035	.005	1.044	.601	.005*	1.06
21	20	.032	.005	1.05	.587	.005*	1.066
22	21	.028	.005	1.06	.546	.004*	1.06
23	22	.025	.005	1.07	.565	.004*	1.07
24	23	.022	.005	1.08	.521	.004*	1.08
25	24	.020	.005	1.088	.503	.004*	1.088
26	25	.018	.005	1.096	.48	.004*	1.096
27	26	.016	.005	1.104	.457	.004*	1.104
28	27	.014	.005	1.1125	.428	.004*	1.1125
29	28	.013	.005	1.1165	.41	.004*	1.1165
30	...	.012	.005	1.1205	.391	.004*	1.1205
...	29	.011	.005	1.1245	.371	.004*	1.1245

* Double silk : 1 mil silk insulation taken equal in weight to 1.25 mil of cotton covering.

67

(b) FORMULÆ FOR WINDING.

Winding of electromagnets. — The
amount of wire which can be wound
upon a spool of given .volume depends
upon the care exercised in the winding
to make it uniform, and upon the amount
of insulation which is placed between the
layers. If V represents the volume of
the spool and V_1 the volume of the wind-
ing, then $V - V_1$, or $\frac{V}{V_1}$, depends upon the
method of winding.

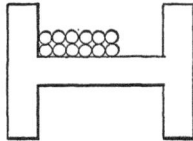

Fig. 15.

Consider the *square* winding, or the
layers superimposed as in Fig. 15, and no
layer insulation. Then

$$\frac{V_1}{V} = \frac{\text{area circle}}{\text{area circumscribed square}}$$

$$= \frac{\pi}{4} = 0.7854.$$

Consider the *conical* winding, or where the convolutions of one layer fit into the recesses of the lower layer, as in Fig. 16.

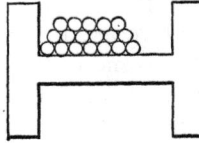

Fig. 16.

Then

$$\frac{V_1}{V} = \frac{\text{area circle}}{\text{area circumscribed hexagon}}$$

$$= \frac{\pi}{\sqrt{12}} = 0.9069.$$

The relation between the volumes occupied in these methods of winding is

$$\frac{\text{conical}}{\text{square}} = \frac{0.9069}{0.7854} = 1.155,$$

i.e., the conical winding contains about 15 per cent. more wire than the square winding.

In practice something between these extreme cases is obtained.

In the above it is assumed that care is exercised in the winding and that all the layers are uniform. In the case of large magnets the layers can be kept regular and the calculated number of turns per layer easily obtained. In the case of magnets where the smaller sizes of wire are used, difficulty is experienced in making the layers uniform. Paper is inserted at intervals between the layers in order to make a smooth bed for the winding. If care is not taken the winding may become lumpy and considerable volume may thus be lost. Where small wire is used the wire volume on a spool may be much less than even the "square winding" would indicate, on account of the paper, irregularities of insulation, lumpiness, etc.

Formulæ for calculation of windings.—The following notation, as applied to Fig. 17, is hereafter employed unless otherwise specified.

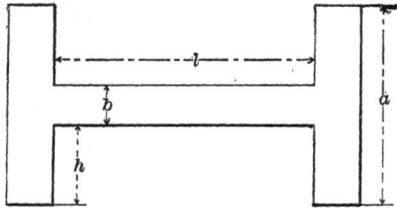

Fig. 17.

D = diameter of covered wire in mils.

d = diameter of bare wire in mils.

t = thickness of insulation on wire in inches, i.e., $\dfrac{D - d}{2}$.

K = ratio of diameter of insulated wire to bare wire.

a, b, h, and l, as in Fig. 17, in inches.

V = volume of winding space in cubic inches.

N = total number of convolutions on spool.

L = total length of wire on spool in feet.

T = number of layers of wire on spool.

n = number of convolutions per linear inch.

$\rho =$ resistance in international ohms of mil-foot of pure copper wire,

$= 10.35$ ohms at $20°C.$ $(68°F.)$.

$R =$ total resistance of coil.

$r =$ resistance per foot of wire in ohms.

$f = \dfrac{1}{r} =$ feet of wire in one ohm.

$l_m =$ mean length of convolution in inches.

The square winding is to be understood unless otherwise specified.

The *total number of convolutions* of wire of given size, filling a given coil space, is

$$N = 1000000 \frac{lh}{D^2} = \frac{500000\, l(a-b)}{D^2}$$

$$= \frac{500000\, l(a-b)}{K^2 d^2}. \qquad (32)$$

The *diameter* of wire necessary to fill a given coil space with a given number of convolutions, is

$$D = \sqrt{\frac{1000000\, lh}{N}} = \sqrt{\frac{500000\, l(a-b)}{N}}.$$

$$(33)$$

or,

$$d = \sqrt{\frac{1000000\ lh}{K^2N}} = \sqrt{\frac{500000\ l(a-b)}{K^2N}}.$$
(34)

The *total length* of wire of given diameter which can be wound in a given coil space is

$$L = \frac{65450\ l(a^2 - b^2)}{D^2}.$$
(35)

From this formula the dimensions of a spool to hold a specified length of wire of given diameter may be determined.

If a and b are known,

$$l = \frac{LD^2}{65450\ (a^2 - b^2)}.$$
(36)

If b and l are known,

$$a = \sqrt{\frac{D^2L + 65450\ lb^2}{65450\ l}}.$$
(37)

If a and l are known,

$$b = \sqrt{\frac{65450\ la^2 - D^2L}{65450\ l}}.$$
(38)

The resistance of a given length of copper wire is

$$R = \frac{L\rho}{d^2}. \qquad (39)$$

Substituting the value of L, as given in Eq. (35), and $\rho = 10.35$.

$$R = \frac{677400(a^2 - b^2)l}{D^2 d^2}. \qquad (40)$$

Now $\qquad (a^2 - b^2)l = \frac{4V}{\pi}.$

Substituting and reducing,

$$R = \frac{862500 \ V}{D^2 d^2}. \qquad (41)$$

If the volume of wire is increased ten per cent to allow for the layers fitting into one another, the above formula (41) becomes

$$R = \frac{948700 \ V}{D^2 d^2}. \qquad (42)$$

Hence, the diameter of wire necessary

to fill a given volume with a given resistance, is

$$d^4 = \frac{948700\ V}{K^2 R}. \qquad (43)$$

Formulæ (41), (42), and (43) are general whatever the shape of the spool, i.e. whether the core is of circular, square, rectangular, elliptical, etc., cross-section.

The next smaller guage number than the diameter corresponding to the formula should be used in order to allow for irregularities in winding and for insulation between the layers.

The deduction of the above formulæ are based upon the assumption that the temperature of the wire is 20°C. (68°F.). If the coil is to be used at some temperature other than 20°C., a new value of R, i.e., R^1 must be used where

$$R^1 = R\ (1\ +\ 0.004\ \theta_c) \qquad (44)$$

where R^1 is the resistance of the heated wire, R the resistance at 20°C. and θ_c the degree rise in temperature above 20°C.

Or, $R^1 = R (1 + 0.0022\ \theta_f)$ (45)

where R is the resistance at 68°F., and θ_f the degree rise in temperature above 68°F.

Table IV in Appendix gives the temperature coefficient for copper wire at various temperatures.

A formula known as Brough's Formula* is often applicable to the calculation of the diameter of wire necessary to give a stated resistance. The constant given for circular core differs from that given in Thompson's Electromagnet, p. 210, in that the diameter of the wire and thickness of insulation are taken in mils instead of in thousandths of an inch.

For circular cores. From (40)

$$R = \frac{677400\ (a^2 - b^2)l}{D^2 d^2}.$$

$$D = d + 2t.$$

$$\therefore\quad d^2(d + 2t)^2 = \frac{677400\ (a^2 - b^2)l}{R}.$$

---

* Journal Society Telegraph Engineers. Vol. V., p. 256.

76

Solving

$$d = \left[\sqrt{\frac{677400\ (a^2 - b^2)l}{R}} + t^2 \right]^{\frac{1}{2}} - t.$$
(44)

For square cores, as in Fig. 18, from (41),

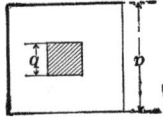

Fig. 18.

$$R = \frac{862500\ l(a^2 - b^2)}{D^2 \times d^2}.$$

$$\therefore d = \left[\sqrt{\frac{862500\ (a^2 - b^2)l}{R}} + t^2 \right]^{\frac{1}{2}} - t.$$
(45)

For rectangular cores, as in Fig. 19,

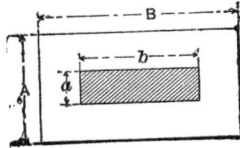

Fig. 19.

$$R = \frac{431250 \,(A - a)\,(A + B + a + b)}{D^2 \times d^2}. \quad (46)$$

$$\therefore \; d =$$

$$\left[\sqrt{\frac{431250\,(A - a)\,(A + B + a + b)}{R}} + t^2 \right]^{\frac{1}{2}}$$
$$- t. \quad (47)$$

For core made up of square and two semicircles, as in Fig. 20,

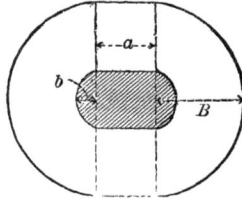

Fig. 20.

$$R = \frac{862500\,(B - b)\,\left[\pi(B + b) + 2a\right]}{D^2 \times d^2}. \quad (48)$$

$$\therefore \; d =$$

$$\left[\sqrt{\frac{862500\,(B - b)\,\left[\pi(B + b) + 2a\right]}{R}} + t^0 \right]^{\frac{1}{2}}$$
$$- t. \quad (49)$$

Relation of ampere-turns to dimensions of coil.—

l_m = mean length of turn.

$\therefore$ $N l_\mathrm{m}$ = total length of wire.

$$R = \frac{N l_\mathrm{m} \rho}{12 d^2}.$$

$$R^1 = R(1 + 0.004 \theta_c) = \frac{N l_\mathrm{m} P}{12 d^2}(1 + 0.004 \theta_c).$$

By Ohm's law $I = \dfrac{E}{R^1}.$

$$NI = \frac{NE}{R^1}.$$

$$\therefore \quad NI = \frac{NE \times 12 d^2}{N l_\mathrm{m} \rho (1 + 0.004 \theta_c)}$$

$$= 1.159 \frac{E d^2}{l_\mathrm{m}(1 + 0.004 \theta_c)}. \tag{50}$$

i.e., the ampere turns are independent of the length of the coil, of the thickness of insulation and of the method of winding; depending upon the diameter of the wire, the mean length of a turn, and the temperature of the coil.

To keep the number of ampere-turns constant in a coil of given volume, d^2 (or the area) of the wire must vary inversely as E.

Given the ampere-turns to find the diameter of wire.—From (50),

$$d^2 = \frac{0.8625\,(NI)\,(1+0.004\,\theta_c)l_m}{E}. \quad (51)$$

If the potential across the terminals of the coil is known, an approximate value of d can be determined from an approximate value of l_m; then d and l_m may be adjusted to bring them within the required limits.

Instead of finding the diameter of the wire it is often more convenient having

Given the ampere-turns to find the feet per ohm.—From (50),

$$\frac{d^2}{\rho} = \frac{(NI)l_m(1+0.004\,\theta_c)}{12\,E},$$

$$R_1 = \frac{L\rho(1+0.004\,\theta_c)}{d^2}.$$

$$\frac{d^2}{\rho(1 + 0.004\ \theta_c)} = \frac{L}{R_1} = f_\theta = \text{feet per ohm}$$

at temperature $(\theta_c + 20)°$.

$$\therefore \qquad f_\theta = \frac{(NI)l_m}{12\ E}. \qquad (52)$$

The feet per ohm must be at the final temperature of the coil. Reference to the resistance tables in the Appendix gives the diameter of wire corresponding to the feet per ohm for various temperatures. The next smaller gauge number should be taken.

Curves which show the relation between the ampere-turns, feet per ohm, mean length of turn, etc., for different voltages, can be easily constructed and furnish a graphical means of determining the constants of a coil.*

Density of current in coil.—The current density in the wire, or the density per square inch, is the current in amperes divided by the cross-sectional area of the

* H. H. Wood. Elec. World, Vol. XXV, 1895, pp. 503, 509.

conductor. The lower the current density the deeper the winding for a given rise of temperature. For coils not exceeding one-half inch thick S. P. Thompson * states that a current density of 3000 amperes per sq. inch can safely be employed. A coil may be designed to give proper current density by proper proportioning of the size of wire.

Heating of magnet coils.—If in a given coil the space taken up by the insulation and the interstices has a fixed ratio to the cross-section of the wire, the magnetizing action and the heating effect remain constant as long as the current remains proportional to the cross-sectional area of the wire; i.e., as long as the current density is constant.

The magnetizing force is proportional to the ampere-turns, i.e., to NI. N varies inversely and I directly as the cross-sectional area of the wire, or,

$$NI \propto \frac{1}{d^2} \times d^2 \; ;$$

* S. P. Thompson " The Electromagnet," 2nd ed. p. 197.

i.e., NI is a constant for a given volume. The heating effect is equal to RI^2. R varies inversely as the square of the cross-sectional area and I directly as the cross-sectional area of the wire, or,

$$RI^2 \propto \frac{1}{d^4} \times d^4 \text{ ;}$$

i.e., RI^2 is a constant for a given volume.

If the current in a given coil is increased the heat effect increases much more rapidly than the magnetizing force, and thus it is the heating effect which in reality determines the dimensions of the wire to be used.

The heat generated in a magnet coil is conducted to the outside of the coil, to the core, and to the heads of the magnet spools. For practical purposes the supposition is generally made that the heat is radiated from the outside or cylindrical surface of the coil. The practical rule determining the area of radiating surface is to allow from 2.0 to 2.5 square inches per watt lost in the coils for a maximum temperature rise of about 75° F.

Esson's * rule for the increase of temperature of magnet coils is

$$\text{Degrees (centigrade)} = 355 \frac{W}{S}, \quad (53)$$

where $W =$ watts lost, and $S =$ superficial area in sq. cms. ; or,

$$\text{Degrees (Fahrenheit)} = 100 \frac{W}{S_1}, \quad (54)$$

where $S_1 =$ superficial area in sq. inches.

Esson states that this rule can be used for all cases in which the coils do not exceed $2\frac{1}{2}''$ or $3''$ in thickness.

It is known that the region of highest temperature lies somewhere in the interior of the coil and that about 65 per cent of the heat developed is radiated from the outer surface of the coil. Carhart's modification of Esson's rule gives †

$$t_0(\text{C.}) = 500 \frac{W}{S}, \quad (55)$$

$$\text{and} \quad t_0(\text{F.}) = 140 \frac{W}{S_1}. \quad (56)$$

---

* Jour. Inst. of Elec. Engrs., Feb. 20, 1890.
† Carhart. Electrical World, 1897, Vol. XXIX, p. 11.

This modification allows about 2 sq. in. per watt for 75° F. rise in temperature.

Carhart has given a provisional empirical formula for the maximum temperature of a coil as

$$T_1 = t + 0.0000445\, d^2 \times D$$

where T_1 is the highest temperature (C.°) when depth of coil is not less than $\frac{3}{4}''$, t is temperature of air (C.°), d is current density, and D is depth of coil (inches).

When a coil carries a large current the wire must be of considerable diameter to avoid undue loss of energy in the coils and excessive heating. On the basis above assumed

$$I^2 R = 0.5\, S_1,$$

where S_1 is in square inches. The maximum current which can then be safely used in a coil of given resistance and surface is

$$I = 0.7 \sqrt{\frac{S_1}{R}}. \tag{57}$$

In the case of a shunt coil, the maximum safe voltage is

$$E = 0.7 \sqrt{S_1 R}. \qquad (58)$$

The energy loss being fixed by the temperature increase, the working resistance of the magnet winding can be obtained by means of Ohm's law.

$$R = \frac{E}{I} = \frac{E \times I}{I^2} = \frac{W}{I^2}, \qquad (59)$$

or $$R = \frac{E}{I} = \frac{E^2}{E \times I} = \frac{E^2}{W}. \qquad (60)$$

The first relation applies to the case of constant current working, while the second applies to constant potential working.

Permissible amperage and permissible depth of winding. — The following formulæ give the permissible amperage for coils of stated dimensions.

Watts $= I^2 R$.

w = watts per square inch of cylindrical surface.

A = area in square inches.

$$I = \sqrt{\frac{12\,d^2 w}{n\mathrm{T}\rho}}. \qquad (61)$$

Take air at 20° C. (68° F.), and a rise of 40° C. (72° F.) gives final temperature of 60° C. (140° F.),

$\rho = 9.612 \times 1.25 = 12.015.$

$w = 0.4$ watt per sq. inch.

Substituting in (61) we have,

$$I = \sqrt{\frac{0.4 d^2}{n\mathrm{T}}}. \qquad (62)*$$

The following tables are derived by means of this formula. For permissible depth of winding 15 per cent is allowed for imbedding of the wires.

---

* Dix, W. S., Elec. Eng., Vol XIV., 1892.

Permissible Amperage and Permissible Depth of Winding for Magnets with Double Cotton-Covered Wire.

Gauge		Diam. Bare, inches.	Circular Mils.	Turns per linear inch.	Layers							
B. & S.	Bir.				1		5		10		20	
					Amp.	Depth.	Amp.	Depth.	Amp.	Depth.	Amp.	Depth.
	2	.284	80656	3.36	97.8	.298	43.7	1.334	31.0	2.63	21.9	5.23
	3	.259	67081	3.66	85.4	.273	38.1	1.222	27.1	2.41	19.1	4.79
2		.2576	66373	3.68	84.6	.2716	37.7	1.216	26.8	2.40	18.9	4.76
	4	.238	56644	3.97	75.4	.252	33.6	1.128	23.9	2.21	16.8	4.42
3		.2294	52634	4.11	71.5	.2434	31.9	1.090	22.7	2.15	16.0	4.27
	5	.22	48400	4.27	67.2	.234	30.0	1.046	21.3	2.07	15.0	4.10
4		.2043	41743	4.58	60.3	.2183	26.9	.980	19.1	1.93	13.5	3.84
	6	.203	41209	4.61	59.8	.217	26.7	.974	18.9	1.92	13.4	3.80
5		.1819	33102	5.10	50.8	.196	22.7	.880	16.1	1.73	11.4	3.44
	7	.18	32400	5.16	50.0	.194	22.3	.871	15.8	1.71	11.2	3.40
	8	.165	27225	5.59	44.1	.179	19.7	.804	13.9	1.58	9.85	3.14
6		.162	26251	5.68	42.9	.176	19.1	.791	13.6	1.55	9.61	3.09
	9	.148	21904	6.17	37.6	.162	16.8	.727	11.9	1.43	8.41	2.84
7		.1443	20817	6.32	36.2	.1583	16.1	.712	11.5	1.40	8.10	2.78
	10	.134	17956	6.76	32.5	.148	14.5	.665	10.3	1.31	7.27	2.59
8		.1285	16510	7.02	30.6	.1425	13.6	.640	9.70	1.26	6.84	2.50
	11	.12	14400	7.46	27.7	.134	12.3	.602	8.78	1.18	6.20	2.35
9		.1144	13004	7.79	25.7	.1284	11.5	.577	8.14	1.13	5.75	2.25

PERMISSIBLE AMPERAGE AND PERMISSIBLE
SINGLE COTTON-

B. & S. Gauge	Bir.	Diameter Bare, inches.	Circular Mils.	Turns per linear inch.	LAYERS.					
					1		2		4	
					Amp.	Depth.	Amp.	Depth.	Amp.	Depth.
..	2	.284	80656	3.44	96.7	.291	68.3	.56	48.4	1.05
..	3	.259	67081	3.76	84.3	.266	59.6	.50	42.2	.96
2	..	.2576	66373	3.78	83.6	.2646	59.1	.50	41.8	.96
..	4	.238	56644	4.08	74.3	.245	52.5	.46	37.1	.89
3	..	.2294	52634	4.23	70.4	.2364	49.8	.44	35.2	.86
..	5	.22	48400	4.41	66.2	.227	46.8	.43	33.1	.82
4	..	.2043	41743	4.73	59.3	.2113	41.9	.40	29.6	.76
..	6	.203	41209	4.76	58.8	.210	39.5	.39	29.9	.76
5	..	.1819	33102	5.29	49.9	.189	35.3	.36	24.9	.68
..	7	.18	32400	5.35	49.1	.187	34.7	.35	24.5	.67
..	8	.165	27225	5.81	43.2	.172	30.5	.32	21.6	.62
6	..	.162	26251	5.92	42.1	.169	29.7	.32	21.1	.61
..	9	.148	21904	6.45	36.8	.155	26.0	.29	18.4	.56
7	..	.1443	20817	6.61	35.4	.1513	25.0	.28	17.7	.55
..	10	.134	17956	7.09	31.7	.141	22.4	.26	15.8	.51
8	..	.1285	16510	7.38	29.9	.1355	21.2	.25	14.9	.49
..	11	.12	14400	7.87	27.0	.127	19.1	.24	13.5	.46
9	..	.1144	13094	8.24	25.1	.1214	17.8	.23	12.57	.44
..	12	.109	11881	8.62	23.4	.116	16.6	.22	11.70	.42
10	..	.1019	10382	9.18	21.3	.1089	15.1	.204	10.63	.39
..	13	.095	9025	9.80	19.2	.102	13.56	.191	9.60	.37
11	..	.0907	8234	10.24	17.9	.0977	12.65	.183	8.95	.35
..	14	.083	6889	11.36	15.37	.088	10.85	.165	7.68	.32
12	..	.0808	6530	11.66	14.93	.0858	10.55	.161	7.46	.31
13	15	.072	5184	12.99	12.61	.077	8.92	.144	6.30	.28
..	16	.065	4225	14.29	10.84	.07	7.66	.131	5.42	.25
14	..	.0641	4107	14.47	10.63	.0691	7.52	.129	5.31	.25
15	..	.0571	3257	16.10	8.98	.0621	6.34	.116	4.49	.224
16	..	.0508	2583	17.92	7.58	.0558	5.36	.105	3.79	.202
17	..	.0453	2048	19.88	6.41	.0503	4.53	.094	3.21	.181
18	..	.0403	1624	22.08	5.41	.0453	3.82	.085	2.71	.164
19	..	.0359	1288	24.45	4.50	.0409	3.24	.077	2.29	.148
20	21	.032	1024	27.03	3.88	.037	2.74	.069	1.94	.137
21	..	.0285	810	29.85	3.28	.0335	2.32	.063	1.64	.121
22	..	.0253	642	33.00	2.79	.0303	1.97	.057	1.39	.109
23	..	.0226	509	36.23	2.37	.0276	1.67	.052	1.184	.100
..	24	.022	484	37.04	2.29	.027	1.616	.051	1.142	.098
24	..	.0201	404	39.84	2.01	.0251	1.420	.047	1.005	.091
25	..	.0179	320	44.64	1.69	.0224	1.193	.042	.845	.081
26	..	.0159	254	49.02	1.436	.0204	1.014	.038	.718	.073
27	..	.0142	201	53.48	1.223	.0187	.865	.035	.611	.068
28	..	.0126	159.8	58.48	1.043	.0171	.737	.032	.521	.062
29	..	.0113	126.7	63.29	.893	.0158	.631	.030	.446	.057
30	..	.01003	100.5	68.97	.762	.0145	.538	.027	.381	.052

DEPTH OF WINDING FOR MAGNETS WITH COVERED WIRE.

LAYERS.

6		8		10		12		16	
Amp.	Depth.	Amp.	Depth.	Amp.	Depth.	Amp.	Depth.	Amp.	Depth.
39.5	1.55	34.2	2.06	30.7	2.57	27.9	3.08	24.2	4.08
34.4	1.42	29.8	1.89	26.7	2.35	24.3	2.81	21.1	3.74
34.1	1.41	29.6	1.87	26.5	2.34	24.1	2.80	20.9	3.72
30.3	1.31	26.1	1.74	23.6	2.16	21.4	2.59	18.6	3.44
28.8	1.26	24.9	1.67	22.3	2.09	20.3	2.50	17.6	3.32
27.0	1.21	23.4	1.61	21.0	2.00	19.1	2.40	16.6	3.19
24.2	1.13	21.0	1.50	18.8	1.86	17.1	2.23	14.8	2.97
24.0	1.12	20.8	1.49	18.6	1.85	16.9	2.22	14.7	2.95
20.4	1.01	17.7	1.34	15.8	1.67	14.4	2.00	12.5	2.86
20.0	1.00	17.3	1.33	15.5	1.65	14.1	1.98	12.27	2.63
17.6	.92	15.3	1.22	13.7	1.52	12.5	1.82	10.80	2.42
17.2	.90	14.9	1.20	13.3	1.49	12.1	1.79	10.52	2.38
15.0	.83	13.0	1.10	11.65	1.37	10.6	1.64	9.20	2.18
14.4	.81	12.5	1.07	11.21	1.33	10.2	1.60	8.85	2.13
12.9	.75	11.2	1.02	10.04	1.24	9.15	1.49	7.92	1.98
12.2	.72	10.6	.96	9.48	1.20	8.63	1.43	7.47	1.90
11.0	.68	9.56	.90	8.56	1.12	7.79	1.34	6.75	1.79
10.2	.65	8.90	.86	7.95	1.07	7.25	1.28	6.27	1.71
9.55	.62	8.28	.82	7.44	1.02	6.75	1.23	5.85	1.63
8.20	.58	7.54	.77	6.75	.96	6.15	1.15	5.32	1.53
7.84	.55	6.80	.72	6.08	.90	5.54	1.08	4.80	1.43
7.31	.57	6.34	.69	5.67	.86	5.17	1.03	4.97	1.37
6.27	.47	5.44	.62	4.87	.78	4.43	.93	3.84	1.24
6.10	.46	5.28	.61	4.73	.76	4.31	.91	3.73	1.21
5.15	.41	4.47	.55	4.00	.68	3.64	.82	3.15	1.08
4.43	.38	3.84	.50	3.44	.62	3.12	.74	2.71	.98
4.34	.37	3.76	.49	3.37	.61	3.07	.73	2.66	.97
3.66	.33	3.18	.44	2.85	.55	2.59	.66	2.24	.87
3.10	.30	2.68	.40	2.40	.49	2.19	.59	1.89	.79
2.62	.27	2.27	.36	2.08	.44	1.85	.53	1.60	.71
2.21	.24	1.91	.32	1.71	.40	1.56	.48	1.35	.64
1.87	.22	1.62	.29	1.46	.36	1.32	.43	1.15	.58
1.58	.198	1.37	.26	1.26	.33	1.12	.39	.970	.52
1.34	.179	1.16	.24	1.04	.30	.946	.35	.820	.47
1.14	.162	.988	.22	.885	.27	.805	.32	.697	.43
.968	.148	.839	.20	.752	.24	.683	.29	.592	.39
.935	.144	.810	.19	.726	.24	.661	.29	.572	.38
.820	.134	.712	.178	.637	.22	.580	.27	.502	.35
.690	.120	.598	.159	.536	.197	.487	.24	.322	.32
.587	.109	.508	.145	.455	.180	.414	.21	.309	.29
.499	.100	.433	.133	.388	.165	.353	.198	.306	.26
.427	.092	.369	.121	.331	.151	.302	.181	.261	.24
.364	.085	.316	.112	.283	.139	.258	.167	.223	.22
.311	.078	.270	.103	.242	.128	.220	.153	.190	.20

Weight of magnet winding in terms of watts lost in winding.—The resistance of a mil-foot of copper wire at a temperature of 20° C. (68° F.) is 10.35 ohms and the weight of 1000 feet is 0.00303 lbs.

$$R = \frac{10.35\,L}{d^2}.$$

The weight of a wire in pounds (W) is

$$W = \frac{L}{1000} \times 0.00303\,d^2.$$

$$\therefore \quad d^2 = \frac{1000\,L}{0.00303\,W}.$$

$$\therefore \quad R = \frac{10.35\,L^2 \times 0.00303}{1000\,W}.$$

$$R = \frac{E}{I} = \frac{watts}{I^2}.$$

$$watts = \frac{10.35\,L^2 \times 0.00303\,I^2}{1000\,W}.$$

or, $$W = \frac{31.4\,L^2 I^2}{(1000)^2\,watts}. \qquad (63)$$

L, the total length of wire, is equal to

the mean turn times the number of turns,
and LI equals the product of the ampere-
turns and the mean length of one turn.

∴ weight of winding (lbs.) =

$$\frac{31.4\left[\dfrac{\text{ampere turns} \times \text{feet per turn}}{1000}\right]^2}{\text{watts lost in winding}}. \quad (64)$$

This formula can be worked backwards
to find the size of wire to be used, but al-
lowance for increased resistance due to
heating must be made if necessary. The
weight of wire can be found that gives a
fixed magnetizing force with a definite
loss of energy in the winding.

*Relations holding between constants
of coils.*—In the following it is considered
that the thickness of the wire insulation
is proportional to the diameter of the wire
and that all coils are uniformly wound.
The results obtained under these consid-
erations are practically but not strictly
correct.

It follows directly that *the weight of
copper required to fill a given coil volume*

is constant whatever the size of the wire used.

From (41) the equation

$$R = \frac{862500\, V}{D^2 d^2} \text{ may be written,}$$

$$R = \frac{A}{d^4}, \text{ where A is a constant as V}$$

is a constant, and $D = Kd$.

It follows directly that *the resistance of a given coil varies inversely as the fourth power of the diameter of the wire used,* or, as the cross-sectional area varies as d^2, *the resistance of a given coil varies inversely as the square of the cross-sectional area of the wire used.*

From equation (32) on page 71,

$$N = \frac{500000\, l(a - b)}{K^2 d^2} = \frac{B}{d^2},$$

where B is a constant for fixed volume. It follows directly that *the number of turns in a fixed volume varies inversely as the square of the diameter of the wire used,* or *inversely as the cross-sectional area of the wire used.*

From the above relation

$$\frac{1}{d^2} = \frac{N}{B}.$$

Hence by substitution,

$$R = \frac{A}{d^4} = \frac{AN^2}{B^2} = cN^2,$$

where c is a constant for fixed volume. It follows directly that *the resistance of a coil varies directly as the square of the number of turns.*

The magnetic effect produced by an electromagnet of given size, shape and construction is proportional to the product of the current into the square root of the resistance of the coil; i.e., magnetic effect or ampere turns varies as $I\sqrt{R}$. Ampere-turns equals NI and as R varies as N^2, N varies as $\sqrt{R}$ and hence NI varies as $I\sqrt{R}$.

If a given coil is wound to a resistance R with a wire of diameter d, what diameter of wire shall be used to rewind to a resistance R_1 ?

$$R \text{ varies as } \frac{1}{d^4}, \text{ i.e., } R = \frac{A}{d^4};$$

$$\text{for same volume, } R_1 = \frac{A}{d_1^4},$$

$$\text{therefore, } \quad \frac{R}{R_1} = \frac{d_1^4}{d^4},$$

$$\text{or,} \qquad d_1^4 = \frac{R}{R_1} d^4. \tag{65}$$

If two coils of the same dimensions are wound with different sized wire, the current must vary with the cross-sectional area of the wire, that is, current density must be the same in each, in order to obtain the same heating effect or the same temperature rise.

$$\text{For} \qquad R = \frac{A}{d^4}.$$

$$\text{Heat loss} = I^2 R = \frac{AI^2}{d^4} = A_1 \frac{I_2}{\text{area}^2}$$

$$= A_2 (\text{density})^2 \cdot$$

Hence, for same energy loss the density must be a constant.

Also,

$$\text{Heat loss} = \frac{E^2}{R} = \frac{E^2 d^4}{A} = \frac{E^2 (\text{area})^2}{A_1}.$$

Hence, for same energy loss, E^2 must vary inversely as the $\overline{\text{area}^2}$, or for same heating effect, the voltage across terminals of coil must vary inversely as the cross-sectional area of the wire used.

Relation of resistance of magnet winding to resistance of external circuit.— An electromagnet is connected to a circuit at the further end of which an E.M.F. is applied; what is the best resistance to give the electromagnet in order to get the maximum magnetic effect?

Let B represent the battery resistance, and R the resistance of the electromagnet, and R_o the resistance of the rest of the circuit. The resultant current is

$$I = \frac{E}{(B + R_o) + R}.$$

The magnetic effect produced by a current I, is proportional to $I \sqrt{R}$, or equal to $KI \sqrt{R}$, where K is some constant. Therefore,

$$\text{magnetic effect} = \frac{KE}{\dfrac{B + R_0}{\sqrt{R}} + \sqrt{R}},$$

and is a maximum when the denominator has a minimum value, i.e., when

$$\frac{B + R_0}{\sqrt{R}} = \sqrt{R}, \text{ or, } (B + R_0) = R, \quad (66)$$

that is, the electromagnet should have a resistance equal to that of the rest of the circuit. It has been proved by Ayrton and Whitehead[*] that in the case of a leaky telegraph line, the resistance of the magnet, A, should equal the apparent resistance of the line tested from the A end, when the further, or B, end is grounded through the apparatus at that end.

It must be noted, however, that while

---

[*] Journal Inst. of Elec. Eng. Vol. XXIII., 1894, p. 327.

the above solution applies to a permanent
state of current, it does not apply to a
variable state of current in general, nota-
bly that in the case of rapid telegraphy.*
While not following the above deduction absolutely in practice, the guiding
factor is that, if working through a low
resistance line, a low resistance battery
and magnet coil should be used ; while
if working through a high resistance line,
a high resistance coil must be used in
connection with a battery of high E. M. F.

* Discussion of Ayrton & Whitehead's Paper by Prof,·
Hughes.
Vaschy, " Electrieité et Magnetisme," Tome II. p. 32.

CHAPTER IV.

In the application of electromagnetism there are many forms of apparatus or mechanisms; some for the consumption of large amounts of energy, such as field magnets, magnets for lifting purposes, etc.; and then those in which only a small amount of energy is consumed, such as those in which a relative motion of parts is produced, as in measuring instruments, relays, bells, telephone receivers, etc.

In all forms of electromagnets the action is based upon the principle of "least reluctance;" that is, the magnetic circuit tends to set itself so that its reluctance reaches a minimum value. The general principle of most electromagnetic mechanisms is the tendency to shorten or close the magnetic circuit by the attraction, or insertion, of an armature when the electromagnet is energized.

Electromagnets may be broadly divided into tractive or portative magnets and attractive magnets according to their function.

TRACTIVE ELECTROMAGNETS.

It may be shown that the force which exists between the pole of an electromagnet and the face of the armature or keeper is

$$P \text{ (dynes)} = \frac{B^2 A}{8\pi}, \qquad (67)$$

where B = induction per sq. cm.

A = area of contact surfaces in $\overline{cm^2}$.

One gramme being equivalent to 981 dynes, approx., (67), reduces to

$$P \text{ (grammes)} = \frac{B^2 A}{8\pi \times 981} = \frac{B^2 A}{24650}. \quad (68)$$

or,

$$P \text{ (kilogrammes)} = \frac{B^2 A}{2465000000}. \qquad (69)$$

$$= \frac{B^2 A}{(5000)^2}, \text{ approximately.}$$

In terms of pounds and sq. cms., (68) becomes

$$P_1 \text{ (lbs.)} = \frac{B^2 A}{11183000}. \qquad (70)$$

In terms of pounds and sq. inches, (68) becomes

$$P_1 \text{ (lbs.)} = \frac{B_1{}^2 A_1}{72134000}. \qquad (71)$$

$$= \frac{B_1{}^2 A_1}{(8500)^2}, \text{ approximately.}$$

This equation gives the weight which a given area will hold up when a certain induction is passing normally through that area.

Let $p_1 =$ pounds per sq. inch; then,

$$p_1 = \frac{P_1}{A_1} = \frac{B_1{}^2}{72134000} = \left(\frac{B_1}{8500}\right)^2, \text{approx.,}$$

which is the pounds per sq. inch that a given induction, B_1, will support.

$$F_1 = B_1 A_1. \qquad \therefore B_1{}^2 A_1 = \frac{F_1{}^2}{A_1}.$$

101

Substituting in (71) and rearranging,

$$A_1 = \frac{F_1^2}{72134000\, P_1},\qquad (72)$$

which is the relation between area, weight to be supported, and total induction.

Also, $\qquad A_1 = \dfrac{72134000\, P_1}{B_1^2}.$

From (71), $\quad B_1^2 = 72134000\, \dfrac{P_1}{A_1}.$

$\therefore\qquad B_1 = 8493\, \sqrt{\dfrac{P_1}{A_1}}$

$$= 8493\, \sqrt{\frac{\text{Pull in pounds}}{\text{Area in sq. inches}}},\qquad (73)$$

which gives the induction per square inch that is necessary to support a given weight over a given area.

Also,

$$B = 1317\, \sqrt{\frac{\text{Pull in pounds}}{\text{Area in sq. inches}}}.\qquad (74)$$

$$B = 4965\, \sqrt{\frac{\text{Pull in kilos}}{\text{Area in sq. inches}}}.\qquad (75)$$

$$B = 157 \sqrt{\frac{\text{Pull in grammes}}{\text{Area in sq. cms.}}}. \quad (76)$$

Equations (30) and (31) differ from the above (76) and (74) by the addition of the term H on the right-hand side of the equation. The reason for this is, that in the method described on p. 57, the core is moved while the coil is fixed, and hence the pull is that due to B—H lines of force.

By means of equations (68) and (71) the following table has been calculated. (See S. P. Thompsou, "The Electro-magnet," p. 119.)

103

MAGNETIZATION AND TRACTION OF ELECTRO-
MAGNETS.

B lines per sq. cm.	B lines per sq. inch.	Grammes per sq. cm.	Kilogs. per sq. cm.	Pounds per sq. inch.
1,000	6,450	40.56	.0456	.577
2,000	12,900	162.3	.1623	2.308
3,000	19,350	365.1	.3651	5.190
4,000	25,800	648.9	.6489	9.228
5,000	32,250	1,014	1.014	14.39
6,000	38,700	1,460	1.460	20.75
7,000	45,150	1,987	1.987	28.26
8,000	51,600	2,596	2.596	36.95
9,000	58,050	3,286	3.286	46.72
10,000	64,500	4,056	4.056	57.68
11,000	70,950	4,907	4.907	69.77
12,000	77,400	5,841	5.841	83.07
13,000	83,850	6,855	6.855	97.47
14,000	90,300	7,550	7.550	113.1
15,000	96,750	9,124	9.124	129.7
16,000	103,200	10,390	10.390	147.7
17,000	109,650	11,720	11.720	166.6
18,000	116,100	13,140	13.140	186.8
19,000	122,550	14,630	14.630	208.1
20,000	129,000	16,230	16.230	230.8

This table is based upon the assumption that there is a uniform distribution of lines of force over the area considered.

On the assumption that there is no magnetic leakage the ampere-turns required for a stated induction are by (22),

$$NI = 0.3132 \ F\Sigma \frac{l}{\mu A}$$

where Σ signifies the sum of the reluctances of the different portions of the magnetic circuit.

Assuming that the iron of the different portions of the circuit is of the same magnetic quality, and that there is no magnetic leakage (22) by the relation $F = B_1 A_1$ reduces to

$$NI = 0.3132 \ \frac{B_1 l_{11}}{\mu} \qquad (77)$$

where l_{11} is the mean total length in inches of the path of the lines of force.

Hence, $\qquad B_1 = \dfrac{\mu NI}{0.3132 \, l_{11}}.$ $\qquad (78)$

Equating (73) and (74), and reducing

$$NI = 2661 \frac{l_{11}}{\mu} \sqrt{\frac{P_1}{A_1}}, \qquad (79)$$

which gives the total number of ampere-turns necessary when the mean length of the magnetic circuit, weight to be supported, and total area are known.

If the dimensions are given in metric measure,

$$NI = 3951 \frac{l}{\mu} \sqrt{\frac{\text{Pull in kilos.}}{\text{Area in sq. cms.}}}. \quad (80)$$

In designing a magnet for tractive purposes, certain assumptions have to be made; first, as to the density of the lines of force. The tractive power increases as B^2 and B should be made as large as possible, about 120,000 lines per sq. in., i.e., for wrought iron. The softest wrought iron of the highest permeability should be used. Having fixed on the value of B, the relation given in (69) and (70), between B and P determines A. From the formula it is seen that a large induction is more essential than a large polar area, as the induction enters the formula as the square while the area enters as the first power. The area of cross-section must, however, be as large as possible. The cross-section may be reduced somewhat at the contact-surfaces; this must not be carried too far, however, as it will tend to increase leakage and reluctance. The reluctance of the circuit must be kept as low as possible, therefore the shape must be compact. The contact surfaces must be carefully faced so that the reluctance of the joints will be as small as possible.

ATTRACTIVE ELECTROMAGNETS.

By an attractive electromagnet is understood one in which a relative motion between the component parts takes place when the coil is energized.

The value of the attraction which exists between any two parts of the magnetic circuit, such as the poles and the armature, depends directly upon the area of the attracting surfaces and the square of the magnetic induction, the separation between poles and armature being constant. If the area and the induction are constant, the attraction varies inversely as the square of the distance separating the attractive surfaces. With a constant M.M.F. the attraction decreases rapidly as the separation increases, on account of the increased separation and the greatly decreased induction caused by the increased reluctance of the magnetic circuit due to the air-gap. For a small separation the attraction on the armature is considerable, and hence, the amount of play of the moving part for a given attraction is small and has to be limited. As the armature is attracted the separation decreases and the attraction increases. The force acting on the armature is thus a variable one.

In order to partly overcome the decrease in magnetic flux due to the increased reluctance, the number of ampere-turns, and hence the M.M.F. may be increased by increasing the number of turns in the coil winding. This increases the length of the core, but the increased reluctance due to the increased length is small compared to the air-gap reluctance.

Numerous plans have been devised to increase the play of the armature and to make the attraction more uniform. Only the general methods can be considered here.

The resultant motion may be increased or equalized by mechanical devices, such as levers, toothed wheels, etc., and variations in attraction may be governed by springs.

The play of the armature may be increased by inserting an iron wedge into the magnetic circuit, the wedge, or armature, being so shaped that there is practically a uniform decrease in the reluctance. The armature may be provided with conical projections which work into conical cavities in the core, or the reverse. In the case of light armatures conical projections on the core may fit into conical shaped holes in the armature.

If a considerable displacement is desired with a small variation in the at-

tractive force, the maximum effect is
produced by the suction of a core into a
solenoid. The field of a solenoid is sen-
sibly uniform in the region of the center
and decreases towards the ends where it
is weakest owing to the demagnetizing
effect of the ends. The magnetic field
of a solenoid is not perfectly uniform.
The magnetic field produced by the induced poles react and have a demagnetizing action as indicated in Fig. 21.

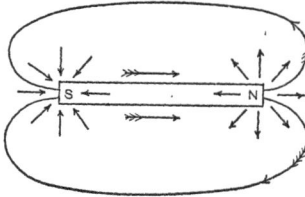

Fig. 21.

The direction of the lines of force produced by the solenoid are indicated by
the feathered arrows, and the plain arrows
indicate the direction of the lines of force
produced by the induced poles, which ra-

diate out from each end. It is seen that
in the interior of the solenoid the two sets
of lines are opposed, thus producing an
unequal distribution of the fields near
the ends. This effect is the more marked
the shorter the coil or the smaller the
ratio of length to diameter. The intro-
duction of an iron core into the solenoid greatly increases this demagnetizing
action.

Consider a core several times longer
than the solenoid. As the core approaches and enters the solenoid the pole
induced at its lower end is attracted and
drawn further in. This action by decreasing the reluctance of the circuit increases the flux and therefore the attraction. When the end of the core approaches the end of the solenoid the
attraction is at a maximum, and as the
end emerges the attraction decreases as
the induced pole enters a region where
the field intensity is diminishing.

If the core has a length the same as
that of the solenoid the action is different,

as the repulsive action of the pole induced at the further end of the core plays an important part. In this case the resultant attraction is a maximum when the end reaches the center of the solenoid, then diminishing and becoming zero when the core is symmetrically placed in the solenoid.

The attraction on the core is somewhat more uniform when the entering end of the core is cone-shaped. The maximum attraction is reached when the apex of the core has emerged some distance out of the coil, but is not as great for the cone-shaped core as for the cylindrical core. By giving the core more complicated forms, the attraction can be made more uniform over a longer range, but with a smaller value than with the cylindrical core.

The suction effect can be much increased by using an iron-clad solenoid, as shown in Fig. 22.

The field at the interior of the solenoid is greatly increased, and the pull is made

112

more uniform. There is no external
field in this case, and the core is not

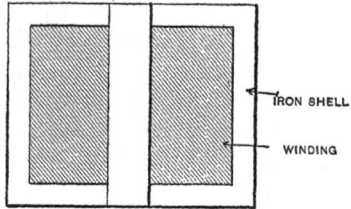

Fig. 22.

attracted until the end is introduced into
one of the openings of the shell. Fig. 23

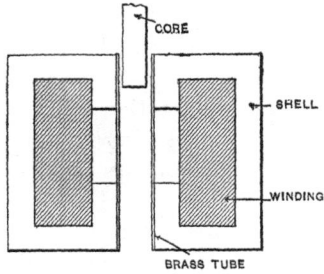

Fig. 23.

shows an extension of the iron-clad type,
there being iron tubes extending into

the interior. The field is thus concentrated into a short space and a very powerful pull is produced on the iron core, or plunger, through a short distance. The range of action can be varied by increasing or decreasing the length of the internal tubes.

Hollow cores, for weak magnetizing forces, act as well as solid ones, but for strong magnetizing forces solid cores are much superior.

Many modifications of the coil and plunger electromagnet have been made, their construction depending upon their use. The coil may be wound in sections, different sections being connected in different circuits as in a shunt circuit, or a series circuit, etc. The coils may be connected differentially, the attraction then depending upon the difference in current strengths. A form known as the differential coil and plunger has had some use in arc lamp regulation, and is shown in Fig. 24.

Here the action of the plunger depends
upon the relative strengths and direction
of the current in the two coils. The mechanical action in B is more independent,
and the magnetization of each plunger

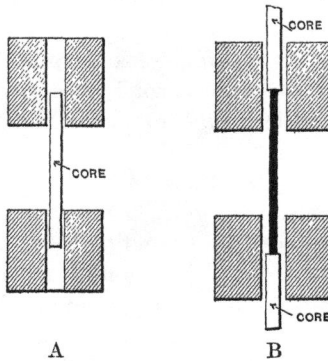

Fig. 24.

is dependent upon the current strength
in its own coil.

Fig. 25 shows the use of a differential
coil and plunger electromagnet in the feed
mechanism of an arc lamp, one coil being
connected in series with the carbon cir-

cuit, and the other circuit in shunt with it.
In this case when the carbons are in contact the series coil becomes energized and
lifts the upper carbon, thus causing some
current to flow through the shunt coil,
and the extent of separation of the car-

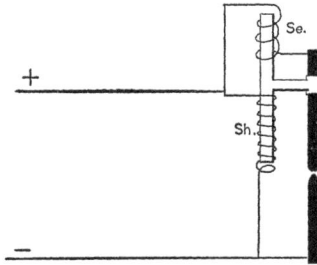

Fig. 25.

bons is dependent upon the relative pull
of the series and shunt coils.

A form of magnet which has an extensive range and a powerful pull over a large
part of that range, combines the construction of both the core electromagnet
and the coil and plunger electromagnet.

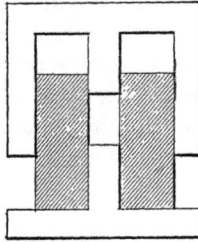

Fig. 26.

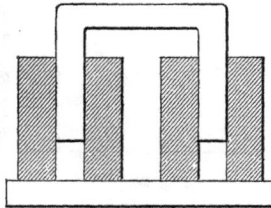

Fig. 27.

Two forms are shown in the Figures 26
and 27.

DESIGN OF TRACTIVE ELECTRO-MAGNETS.*

An electromagnet performs work when a loaded armature, placed at a given distance from its poles, is attracted and drawn up to the poles. Work done by an electromagnet may be defined as the product of the initial pull exerted upon the armature and the travel of the armature. It may be shown that this product is a maximum when the armature is placed at such a distance from the poles that the ratio of the rate of change of the magnetic flux in the iron to the rate of change of the magnetizing force producing it is equal to the permeance of the working air-gaps.

Let M be the m.m.f. acting on all the circuit except the working air-gap.

---

* W. E. Goldsborough, Elec. World, Vol. XXXVI, p. 125, July 28, 1900. The following pages (117-121) are taken almost literally from this article. For further details, reference should be made to this article. Further see E. R. Carichoff, Elec. World, Vol. XXIII, p. 113 and p. 212. 1894. C. T. Hutchinson, Elec. World, Vol. XXIII, p. 243, 1894.

Let M^1 be the m.m.f. acting on the working air-gap.

Then $M + M^1$ is a constant, and equal to the applied m.m.f.

Let F_c be the total magnetic flux in the core of the exciting coil.

Let F be the total magnetic flux in the working air-gap.

Let A^1 be the working area of the air-gap.

Let L be the length of the working air-gap.

Then
$$\frac{dF}{dM} = \frac{F}{M^1} = \frac{A^1}{L}, \qquad (81)$$

The meaning of this relation may be better understood by a study of Fig. 28.

The m.m.f. acting on the magnetic circuit is a constant, and the relative values of M and M^1, i.e., of the m.m.f.'s required to force the flux F, through the iron and the air-gap is dependent upon the value of L. The greater L, the greater the value of M; the smaller L, the greater the value of M^1.

119

The curve OSJ is plotted from the
values of the total flux and the corresponding values of the m.m.f. (M) im-
pressed upon all of the magnetic circuit
outside of the working air-gap. The
curve KTI is plotted from the values of

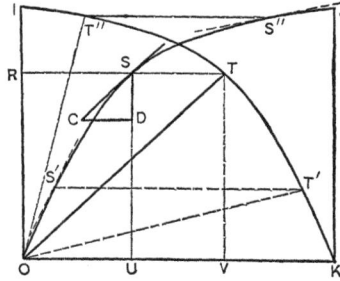

Fig. 28.

the total flux in the working air-gap and
the corresponding values of the m.m.f.
(M') required to force this flux across the
working air-gap. Since M + M¹ is a
constant the sum of the abscissae of the
two curves, taken for successive values of
F, will give the straight line K J.

The slant of the tangent to the curve OSJ at any point, as S, will graphically represent the ratio $\dfrac{d\text{F}}{d\text{M}}$, since SD and CD are respectively proportional to dF and dM. Taking the point T on the curve KTI, that has the same ordinate as the point S, and connect T and O, the line OT will represent the ratio $\dfrac{\text{F}}{\text{M}^1}$, since TV and OV are respectively equal to F and M^1.

The excitation remaining constant, as the armature moves up to its seat, the points S and T move along the curves OSJ and KTI, towards J and I, respectively, and when they arrive at such positions that the tangent to the curve at S is parallel to a line connecting O and T the condition expressed by equation (81) is fulfilled, and the armature is in that position which admits of the magnet doing the maximum amount of work. The points S and T will always be opposite one another, and have the same ordinate values.

When the armature is upon its seat, i. e. makes contact with the pole pieces, the points S and T will be at J and I respectively, and the magnet will be exerting its maximum grip upon the plunger or armature. Under these conditions the magnet is capable of sustaining a large weight, and essentially presents all the features of the ordinary portative electromagnet.

When the armature is removed to a great distance, the points S and T approach the positions O and K, respectively, but owing to the fact that the mean density in the working air-gap can never be reduced to zero, they will stop short of the limits, in some position where but a feeble pull is exerted upon the armature. The action of the magnet upon its armature for points on the curves below the points S and T, as shown, is analogous to the action of an iron-clad solenoid upon its plunger, especially when the electro-

magnet is of the iron-clad type, and the
armature or plunger is conical shaped
and works into a conical shaped recess.

ELECTROMAGNETS FOR PRODUCING STRONG FIELDS.

In electromagnets for producing strong
fields the magnetic circuit should be rel-
atively short and should be mechanically
strong so as to prevent springing by the
attraction of the poles under the influence
of strong fields.

To produce a strong field it is first
necessary to obtain as strong a field as
possible by the use of a large number of
ampere-turns, and then to use specially
formed pole-pieces in order to concentrate
the field. In order to obtain intense field
pointed pole-pieces have long been used.
Theory indicates, and experiment con-
firms, that a conical pole-piece of 120°
aperture gives the maximum field, of say
40,000 lines per sq. cm. over an extent of
several square millimeters. The radius

of the pole face should be small, about
1.5 m.m.(¹)

POLARIZED MECHANISMS.

A form of mechanism in which a permanent magnet is used in addition to an electromagnet has extensive applications in practice, and is known as a polarized mechanism. Such mechanisms are used in telegraphy, as the polarized relay; in telephony, as the polarized bell and telephone receiver; and in many forms of signaling apparatus. This form of mechanism possesses several marked properties as follows.

First. Bilateral Action.—In the ordinary electromagnet, the armature is attracted whatever the direction of the current around the core. If the armature is a permanent, or an induced magnet, the armature is either attracted or repelled according to the direction of cur-

---

(1) For further details see Du Bois, loc. cit., p. 258.

rent in the coils, as shown in Fig. 29,
where the armature is shown pivoted at
one end.   If the direction of current is
as shown the armature is attracted to the
left, and if the direction of current is in

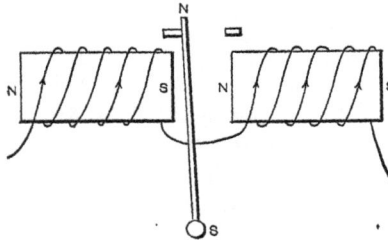

Fig. 29.

the opposite direction it is attracted to
the right.   The direction of deflection is
dependent upon the direction of cur- .
rent flow in the coils.   If the relation
of armature and coils is as shown in Fig.
30, where the polarized armature is pivoted at the centre, so that one end is over
each core, the right end is repelled and
the left end attracted.   Reversing the

direction of current, reverses the attraction and repulsion. If the armature in either case carries a hammer it may be arranged to vibrate against a bell.

The armature in place of being polarized, may be of soft iron, while the cores are polarized. Then one end of the ar-

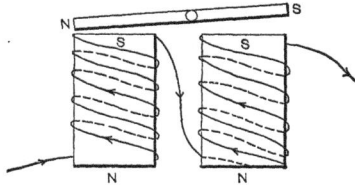

Fig. 30.

mature is attracted on account of the excess pull which one core possesses over the other, when the coils are energized. Reversing the direction of current, reverses the armature attraction. In case of a single polarized core, as in a telephone receiver, the armature has a zero position, or one of rest, and either approaches or recedes from the pole piece according to the direction of current in the coil.

In virtue of this bilateral action an electromagnet is responsive to an alternating current if the inertia of the moving parts is not too great.

Second. Rapidity and sensitiveness of action.—Polarized mechanisms can be worked with great rapidity, as the attractive force of the permanent magnet may be just balanced by a spring, so that with the least addition to the field strength the balance is overcome and the armature moved. In some of the most rapid forms of mechanism, such as the printing telegraph, etc., polarized mechanisms are used.

Greater sensitiveness can also be obtained with weak currents or with small changes of currents. According to the law of traction the force acting on the armature is

$$F = \frac{SB^2}{8\pi}. \qquad (82)$$

for a given separation. For a given
separation and area it is proportional to B^2.
If the induction is increased by an amount
ΔB the pull becomes proportional to
$(B + \Delta B)^2$ and the increased pull is pro-
portional to $2B\Delta B + \overline{\Delta B}^2$. As $\overline{\Delta B}^2$ is
small compared to $2B\Delta B$, the change in
pull is proportional to $2B\Delta B$, that is, the
change in pull is proportional to the
initial induction as well as to the change
in induction, and the greater the initial
induction the greater the change in pull
due to small change in the induction.

In a telephone receiver the permanent
magnet is necessary to keep the pitch of
the emitted tone the same as that of the
transmitted tone.

*Electromagnets for use with currents
of brief duration.*—In telegraphy, use of
the chronograph, signaling apparatus,
etc., it is essential that the electromagnet
should respond to currents of very brief
duration. This is in part accomplished
by making the movable parts light and of

small inertia, and by placing the armature in a biased position.

The rapidity of working is further dependent upon a property of the electric circuit known as its self-induction or inductance. The magnetic field set up about a conductor during a change of current flow in the conductor, generates an E.M.F. which is in opposition to the E.M.F. producing the current. The coefficient of self-induction may be defined as *the ratio of the number of lines of force linked with a circuit to the current producing them.* If F is the flux produced by unit current, N the number of turns, or convolutions, in the winding, then

$$L = FN. \qquad (83)$$

The *henry* is the practical unit of inductance and is 10^9 times the C.G.S. unit of inductance.

Then $$L = \frac{FN}{10^9} \text{ henrys.} \qquad (84)$$

In any magnetic circuit, where I is in amperes,

$$F = \frac{\frac{4\pi}{10} NI}{\Sigma \frac{l}{A\mu}}. \tag{85}$$

Therefore

$$L = \frac{4\pi N^2 \Sigma \frac{A\mu}{l}}{10^9} \text{ henrys.} \tag{86}$$

The inductance varies directly as the square of the number of turns in the winding and directly as the permeability of the magnetic circuit. As the permeability varies with the magnetizing current it is impossible to calculate exactly the value of L from the dimensions of the circuit. If the working value of I is known, a value of μ may be approximated, and comparative values of L may be thus obtained. For exact work L must be measured under working conditions.

When a circuit containing inductance is closed the current does not instantly rise to its full value, but lags behind as if possessing inertia. The law which expresses the strength of current in a circuit at a time t after the closing of the circuit is

$$i_t = I\left(1 - e^{-\frac{Rt}{L}}\right) = \frac{E}{R}\left(1 - e^{-\frac{Rt}{L}}\right) \quad (87)$$

where i_t stands for the current strength at time t (seconds), I, E, R and L have their usual meaning, and e is the number 2.7183. The ratio $\frac{L}{R}$ is known as the time constant of the circuit, and is usually denoted by the letter T. Then

$$i_t = I\left(1 - e^{-\frac{t}{T}}\right) = \frac{E}{R}\left(1 - e^{-\frac{t}{T}}\right). \quad (88)$$

At the instant of closing the circuit $t = o$ and therefore $i_t = o$. When $t = T$, $i_t = I\left(\frac{e-1}{e}\right) = 0.632I$. The time-constant of a circuit can then be defined as the time from closing the circuit in which the current rises to 0.63 times its maximum, or Ohm's law value.

The time-constant may be reduced by decreasing L, or increasing R, or both. In a coil of given volume, the inductance and resistance are both proportional to the square of the number of turns, so that the time-constant of a given coil is not changed by rewinding it. In considering a special problem the time-constant of the circuit as a whole must be considered.

The time-constant of a circuit may be changed by placing the coils of an electro-magnet in parallel instead of in series. Consider the case of a 150 ohm Morse relay connected at the end of an overhead line 200 miles long. The inductance of a single copper wire 0.104″ in diameter, suspended 20 ft. above the ground is about 2.5 milhenrys per mile, and the resistance about 5.1 ohms per mile. The resistance of the Morse relay is about 150 ohms, and it has a working inductance of about 4 henrys. The time-constant, i.e. $\frac{L}{R}$, for the whole circuit is 0.00375 seconds. If the current in the line is considered constant, placing the two windings of the relay in parallel has the resultant effect of a relay whose resistance and inductance are one-fourth of the series values. Putting the coils of the relay in parallel makes the circuit resistance 1087 ohms, the inductance 1.5 henrys, and the time-constant becomes 0.00138 seconds. The time-constant of the circuit with the parallel arrangement of coils is about one-third of that with the series arrangement.

Curves showing the rate of rise of current in the above two circuits are shown in Fig. 31, where the abscissae are ten-thousandths of a second, and the ordinates per cent. of steady value, or $\frac{E}{R}$ value, of current. The decrease in time-constant due to parallel arrangement is marked.

The magnetizing power, or ampere-turns, of the parallel arrangement is one-half of that of the series arrangement, i.e.,

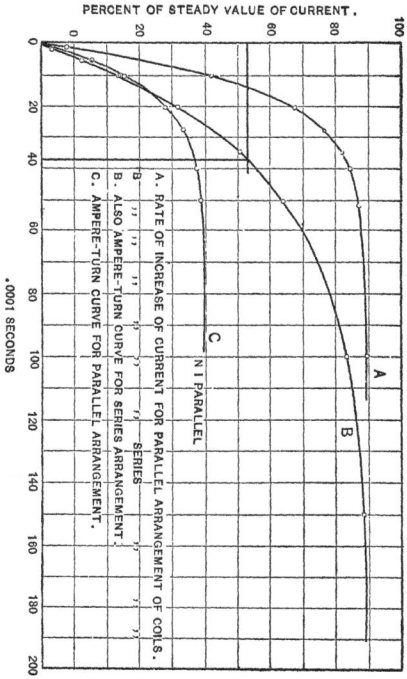

PERCENT OF STEADY VALUE OF CURRENT.

.0001 SECONDS

A. RATE OF INCREASE OF CURRENT FOR PARALLEL ARRANGEMENT OF COILS.
B. " " " " " " SERIES "
B. ALSO AMPERE-TURN CURVE FOR SERIES ARRANGEMENT.
C. " " " " "
C. AMPERE-TURN CURVE FOR PARALLEL ARRANGEMENT.

N I PARALLEL

Fig. 31.

in a constant current system. The magnetizing force being proportional to the current, curve I also gives the rate of increase of magnetizing force in the series-circuit, while curve III gives the rate of increase of magnetizing force for the parallel arrangement of coils. It is obvious that for very brief currents the magnetizing force is the greater for the parallel arrangement of coils.

In the consideration of quick-working electromagnets the resistance and inductance of the entire circuit must be considered in order to determine the best arrangement of terminal apparatus. The electro-static capacity of the line should be considered in very rapid telegraph work, but it makes the complete discussion of the problem very complicated.

CHAPTER V.

ALTERNATE CURRENT ELECTRO-MAGNETS.

BY an alternating current is meant one which is periodically changing its direction of flow. From the consideration of the simplest forms of generator it may be shown that the E.M.F. varies harmonically; and it will be considered that the current varies according to a simple sine law as shown in Fig. 32.

Fig. 32.

Instead of referring to the mean or maximum value of an alternating current or E.M.F., the *effective* or *virtual* value is usually employed. The virtual value of an alternating current is the value of an unvarying or direct current which would produce the same heating effect, and is equal to the square root of the mean square of the instantaneous values.

$$\text{Virtual value} = \frac{\text{max. value}}{\sqrt{2}} = 0.707 \text{ max.}$$
value. (89)

Max. value $= \sqrt{2}$ virtual value $= 1.414$ virtual value. (90)

When a simple harmonic E.M.F., E sin pt, is applied to a circuit of resistance, R, and inductance, L, the relation between E.M.F. and current is

$$I = \frac{E}{\sqrt{R^2 + p^2 L^2}} \sin (pt - \theta), \quad (91)$$

where $p = 2\pi n$, and n is the frequency or number of complete cycles per second, and $\theta = $ arc tan. $\frac{Lp}{R}$, i.e., the angle by which the current lags behind the E.M.F. in an inductive circuit. The quantity $(R^2 + p^2 L^2)^{\frac{1}{2}}$ is termed the impedance of the circuit. The quantity pL is called the reactance. Hence,

$$\overline{\text{Impedance}^2} = \overline{\text{Resistance}^2} + \overline{\text{Reactance}^2}.$$
$$(92)$$

In alternating current work it is necessary to know the inductance of the circuit, and this may be obtained by direct measurement, or, in the case of a closed magnetic circuit by the formula

$$L = \frac{4\pi N^2 A \mu}{10^9 l} \text{ henrys}, \qquad (93)$$

where A is the cross-sectional area of the core in sq. cms., and l is the mean perimeter of the core in cms.

Hysteresis Loss.—It has been stated on page 42 that when the magnetic state is passing through a cycle of changes, energy is lost, due to hysteresis. This lost energy manifests itself in the form of heat. For alternate current work this loss of energy may be large, and undue heating of the core may result. The energy lost in hysteresis is given by the relation

Energy lost (watts) =

$$\frac{\text{volume in cu. cms.}}{10^7}f\eta B^{1.6}_{max.} \quad (94)$$

The following table gives the value of the hysteresis coefficient, η, for several grades of iron.

Grade of Iron.	Coefficient, η.
Very soft iron wire,	0.0020
Very thin soft sheet iron,	0.0024
Sheet iron,	0.0045
Soft annealed cast steel,	0.0080
Cast steel,	0.0120
Cast iron,	0.0162

The purer the iron the smaller the value of η.

Eddy Currents.—When a conducting body is exposed to an alternating magnetic field, electric currents are set up in the mass of the metal, due to the cutting of the lines of force. Such currents are known as *eddy* or *Foucault* currents. Eddy currents are a purely electrical phenomena, while hysteresis loss is a magnetic phenomena. Eddy currents produce a loss of energy and are dependent upon the shape of the metal and its degree of subdivision.

Power lost in eddy currents =

$$\frac{\text{Vol. in cu. cms.}}{10^7} f^2 E B^2 \quad (95)$$

where E is the coefficient of eddy currents, and depends upon the material and shape of the parts of the magnetic circuit.

If the magnetic core is made up of laminated or sheet iron

$$\text{Power lost (watts)} = 1.645 \, d^2 f^2 B^2 \times 10^{-11} \quad (96)^1$$

per unit volume, where d is the thickness of the iron sheet in cms.

If the magnetic core is made up of iron wire,

$$\text{Power lost (watts)} = 0.617 \, d_1^2 f^2 B^2 \times 10^{-11} \quad (97)^1$$

per unit volume, where d_1 is the diameter of the wire in cms.

(1) Steinmetz. "Alternating Current Phenomena." 3d ed. pp. 132, 135.

The energy loss by eddy currents is decreased by increasing the degree of subdivision of the core.

For straight (or bar) electromagnets fine iron wire may be used for the core, while in horse-shoe magnets the core may be built up of iron stampings. If stampings are used the eddy currents may be further decreased by covering the sheets with shellac, or by interposing layers of insulating material, such as paper, mica, etc., and thus decreasing the electrical conductivity of the core. The oxide on the surface of the iron is usually relied upon for insulating purposes.

If the current flowing through the coil of an electromagnet varies periodically the resultant magnetic flux also varies periodically. The armature is then acted upon by a varying force which passes successively from zero to a maximum, and then back to zero. For one complete cycle of current or magnetic flux the armature has two maximums of attraction. For a current of low frequency, the armature will vibrate with a frequency double that of the current. As the current frequency increases the vibrations of the armature will be less manifest, and it will chatter against its stop, and finally, for higher frequencies the armature will be attracted without any chattering. The point at which the vibration and chattering cease depends largely upon the inertia of the armature.

The polarized type of electromagnet can be made to respond to a very small alternating current when the armature is placed in a biased position.

When current flows in two parallel conductors in the same direction there is an attraction between the two conductors, and a repulsion if the currents are flowing in opposite directions.

Consider two conductors conveying alternating currents which are in phase with one another, as indicated by Fig. 33.

There is a varying attraction between the conductors, which has maximum values at b and d. When the current

142

frequency is high the wires do not vibrate
but are urged towards one another.

Consider two conductors conveying alternating currents which are of opposite
phase, i.e., 180° apart, as indicated by

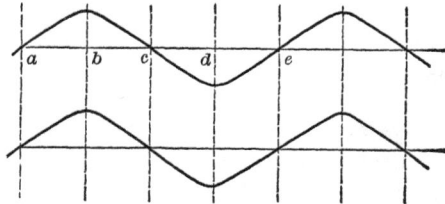

Fig. 33.

Fig. 34. Here there is a varying repulsion between the conductors which has
maximum values at *b* and *d*. When the
frequency is high the wires do not vibrate,
but are urged away from one another.

Consider the intermediate case where
the currents in the two conductors are in
quadrature with each other, i.e., 90° apart
in phase, as indicated by Fig. 35.

At *a*, *b*, *c*, *d* and *e* there are no forces

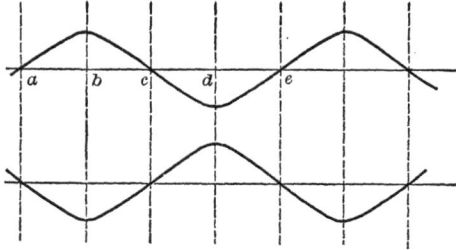

Fig. 34.

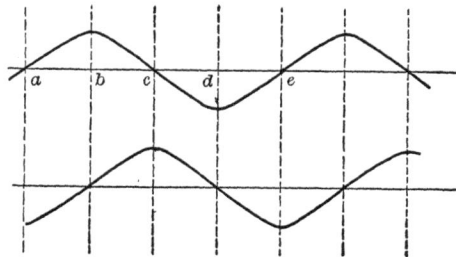

Fig. 35.

acting on the conductors, as one current
has a zero value. From a to b there is
repulsion, from b to c, attraction, from c
to d repulsion, and from d to e attraction.
At low frequencies the conductors vibrate
with double the frequency of the current,
but at high current frequencies there is
no tendency to vibration.

If the current in the conductors are
out of step, and the difference is less than,
or greater than, 90°, the attractions or
repulsions preponderate, and there is a
resultant motion; it being one of attraction if the phase difference is less than
90°, and one of repulsion if the phase
difference is greater than 90°.

Consider a disc of copper which is held
in front of the pole of an alternating current electromagnet. The magnetic field
which cuts the disc is alternating between
two equal and opposite values. An alternating E.M.F. is generated in the disc,
and hence there is an alternating current
flowing around the disc, and consequently
there is a force acting between the core

and the disc.　Consider the upper half of
Fig. 36.

Let the chain dotted curve, A B C D E,
represent the variation of the magnetic

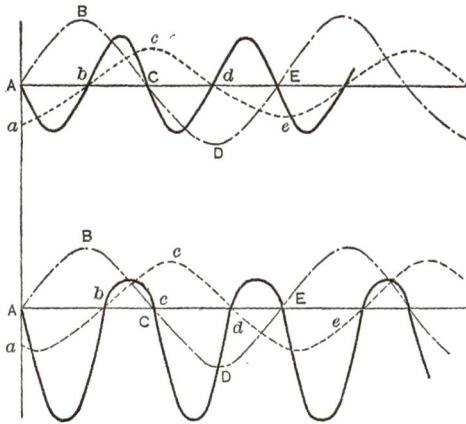

Fig. 36.

field produced by the alternating current.
The E.M.F., which is induced in the
disc is proportional to the rate of change
of the magnetic flux, and is represented

by the dotted curve *a b c d e*. If the disc
was devoid of inductance the current in
it could also be represented by the curve
a b c d e. As the curves A B C D E and
a b c d e are in quadrature there would be
no resultant force acting on the disc; and
the full line shows the relative attractions
and repulsions and their equality.

The disc, however, possesses a certain
amount of inductance, and the current lags
behind the induced E.M.F., as shown by
the dotted line in the lower part of Fig. 36.
The disc is now subjected to alternate
attractions and repulsions, but the repulsions preponderate and there is a resultant
repulsion of the disc from the core of the
electromagnet. The full line in the lower
part of Fig. 36 shows the relative value
of the attractions and repulsions.

If the disc is pivoted so as to rotate it
will rotate into a position with its plane
at right angles to the core. In this
position the induction and hence the eddy
currents are at a minimum. This prop-

erty of the repulsion of discs or masses of metal may be made use of in electro-magnets where repulsion of the armature is desired.

The disc should be made of copper, so that its resistance will be low, and the eddy currents as large as possible

APPENDIX.

I.—TABLES GIVING RELATIONS BETWEEN
U. S. AND METRIC SYSTEMS.

1 Meter = 39.37 Inches.

		LONG MEASURE.		
No.	64th of an Inch to Millimeters.	Millimeters to 64to of an Inch.	Inches to Centimeters.	Centimeters to Inches.
1	0.397	2.520	2.54	0.394
2	0.794	5.039	5.08	0.787
3	1.191	7.559	7.62	1.181
4	1.587	10.079	10.16	1.575
5	1.984	12.598	12.70	1.968
6	2.381	15.118	15.24	2.362
7	2.778	17.638	17.78	2.756
8	3.175	20.157	20.32	3.150
9	3.572	22.677	22.86	3.543

	SQUARE MEASURE.		CUBIC MEASURE,	
No.	Sq. Inches to Square Centimeters	Square Centimeters to Square Inches.	Cubic Inches to Cubic Centimeters.	Cubic Centimeters to Cubic Inches.
1	6.452	0.155	16.393	0.061
2	12.903	0.310	32.787	0.122
3	19.355	0.465	49.180	0.183
4	25.806	0.620	65.574	0.244
5	32.258	0.775	81.947	0.305
6	38.710	0.930	98.361	0.366
7	45.161	1.085	114.75	0.427
8	51.613	1.240	131.15	0.488
9	58.065	1.395	147.54	0.549

II.—Properties of Copper Wire.

Brown and Sharpe Gauge.

Gauge Number.	Diameter in Mils.	Area in Circular Mils. C.M.=d^2	Weight. Pounds per Foot. 20° C. = 68° F.	Length. Feet per Ohm. 20° C. = 68° F.	Resist'nce Ohms per Foot. 20° C. = 68° F.
1	289.3	83,690	0.2533	8,083	0.0001237
2	257.6	66,370	0.2009	6,410	0.0001560
3	229.4	52,630	0.1593	5,084	0.0001967
4	204.3	41,740	0.1264	4,031	0.0002480
5	181.9	33,100	0.1002	3,197	0.0003128
6	162.0	26,250	0.07946	2,535	0.0003944
7	144.3	20,820	0.06302	2,011	0.0004973
8	128.5	16,510	0.04998	1,595	0.0006271
9	114.4	13,090	0.03963	1,265	0.0007908
10	101.9	10,380	0.03143	1,003	0.0009972
11	90.74	8,234	0.02493	795.3	0.001257
12	80.81	6,530	0.01977	630.7	0.001586
13	71.96	5,178	0.01568	500.1	0.001999
14	64.08	4,107	0.01243	396.6	0.002521
15	57.07	3,257	0.009858	314.5	0.003179
16	50.82	2,583	0.007818	249.4	0.004009
17	45.26	2,048	0.006200	197.8	0.005055
18	40.30	1,624	0.004917	156.9	0.006374
19	35.89	1,288	0.003899	124.4	0.008038
20	31.96	1,022	0.003092	98.66	0.01014
21	28.46	810.1	0.002452	78.24	0.01278
22	25.35	642.4	0.001945	62.05	0.01612
23	22.57	509.5	0.001542	49.21	0.02032
24	20.10	404.0	0.001223	39.02	0.02563
25	17.90	320.4	0.0009699	30.95	0.03231
26	15.94	254.1	0.0007692	24.54	0.04075
27	14.20	201.5	0.0006100	19.46	0.05138
28	12.64	159.8	0.0004837	15.43	0.06479
29	11.26	126.7	0.0003836	12.24	0.08170
30	10.03	100.5	0.0003042	9.707	0.1030
31	8.928	79.70	0.0002413	7.698	0.1299
32	7.950	63.21	0.0001913	6.105	0.1638
33	7.080	50.13	0.0001517	4.841	0.2066
34	6.305	39.75	0.0001203	3.839	0.2605
35	5.615	31.52	0.00009543	3.045	0.3284
36	5.000	25.00	0.00007568	2.414	0.4142

Compiled from " Copper Wire Tables " of Am. Inst. of Elec. Engrs.

III.—PROPERTIES OF COPPER WIRE.

B. W. GAUGE.

Gauge Number.	Diameter in Mils.	Area in Circular Mils. C. M. $= d^2$	Weight. Pounds per Foot. $20°$ C. $=$ $68°$ F,	Length. Feet per Ohm. $20°$ C. $=$ $68°$ F.	Resist'nce Ohms per Foot. $20°$ C. $=$ $68°$ F.
1	300.0	90000.	0.2724	8,692.	0.0001150
2	284.0	80660.	0.2441	7,790.	0.0001284
3	259.0	67080.	0.2031	6,479.	0.0001543
4	238.0	56640.	0.1715	5,471.	0.0001828
5	220.0	48400.	0.1465	4,675.	0.0002139
6	203.0	41210.	0.1247	3,980.	0.0002513
7	180.0	32400.	0.09808	3,129.	0.0003196
8	165.0	27230.	0.08241	2,629.	0.0003803
9	148.0	21900.	0.06630	2,116.	0.0004727
10	134.0	17960.	0.05435	1,734.	0.0005766
11	120.0	14400.	0.04359	1,391.	0.0007190
12	109.0	11880.	0.03596	1,147.	0.0008715
13	95.0	9025.	0.02732	871.7	0.001147
14	83.0	6889.	0.02085	665.4	0.001503
15	72.0	5184.	0.01569	500.7	0.001997
16	65.0	4225.	0.01279	408.1	0.002451
17	58.0	3364.	0.01018	324.9	0.003078
18	49.0	2401.	0.007268	231.9	0.004312
19	42.0	1764.	0.005340	170.4	0.005870
20	35.0	1225.	0.003708	118.3	0.008452
21	32.0	1024.	0.003100	98.90	0.01011
22	28.0	784.	0.002373	75.72	0.01321
23	25.0	625.	0.001892	60.36	0.01657
24	22.0	484.	0.001465	46.75	0.02139
25	20.0	400.	0.001211	38.63	0.02588
26	18.0	324.	0.0009808	31.29	0.03196
27	16.0	256.	0.0007749	24.73	0.04045
28	14.0	196.	0.0005933	18.93	0.05283
29	13.0	169.	0.0005116	16.32	0.06127
30	12.0	144.	0.0004359	13.91	0.07190
31	10.0	100.	0.0003027	9.658	0.1035
32	9.0	81.	0.0002452	7.823	0.1278
33	8.0	64.	0.0001937	6.181	0.1618
34	7.0	49.	0.0001483	4.733	0.2113
35	5.0	25.	0.0007568	2.414	0.4142
36	4.0	16.	0.0004843	1.545	0.6471

154

IV.—Temperature Coefficients.

Temperature Degrees.		Temperature Coefficient of Resistance.	
Centigrade.	Fahrenheit.	0°C. = 32° F. as Standard.	20° C. = 68° F. as Standard.
0	32	1.0000	......
5	41	1.0195	......
10	50	1.0393	......
15	59	1.0594	......
20	68	1.0797	1.0000
25	77	1.1003	1.0191
30	86	1.1210	1.0383
35	95	1.1422	1.0579
40	104	1 1633	1.0774
45	113	1.1848	1.0973
50	122	1.2063	1.1172
55	131	1.2282	1.1375
60	140	1.2500	1.1577
65	149	1.2716	1.1777
70	158	1.2933	1.1978
80	176	1.3368	1.2324
90	194	1.3800	1.2781
100	212	1.4223	1.3173

www.ingramcontent.com/pod-product-compliance
Lightning Source LLC
Chambersburg PA
CBHW031402180326
41458CB00043B/6573/J